SOME UNUSUAL ENGINES

L J K SETRIGHT

SOME UNUSUAL ENGINES

Illustrations Selected by

MICHAEL WORTHINGTON-WILLIAMS

Published by

MECHANICAL ENGINEERING PUBLICATIONS LIMITED

for The Institution of Mechanical Engineers

LONDON

First published 1975

ISBN 0 85298 208 9

Published by The Institution of Mechanical Engineers

Designed by Derek Morrison

Set by Page Brothers (Norwich) Ltd, Norwich, Norfolk
Printed by photolitho in Great Britain by Perivan Press WLP Group
Bound by G and J Kitcat Ltd, London, England

Made and printed in Great Britain

Contents

Introduction: Engines that might have been and engines that might well be....

Orthodoxy is my doxy, as Bishop William Warburton explained to Lord Sandwich on an occasion I never tire of quoting; *heterodoxy is another man's doxy*. This Attic wit appears not to have been lost on the internal combustion engine industry, the successful portions of which conduct themselves as though nothing would ever persuade them to acquire another man's tastes. There is a terrible inertia among the designers – or is it the builders? – of engines, a woebegone affection for their own familiar ruts that drives them to a continuing fear of the devil they do not know and an unwholesome regard for the devil they do.

Nevertheless history is not without its moments of relief. There have been many extremely unusual engines conceived, designed, and occasionally built, by would-be iconoclasts. Precious few of them have ever enjoyed any kind of commercial success, even those which were proved to work well. The reason is not far to seek: if there be anybody more conservative than an engine manufacturer, it is his customer. Traditions exist to be followed, and it is only where there are no traditions, and where it is accepted that an engine has some particular duty to do but is otherwise likely to be covered up and ignored by everybody except the minions delegated to look after it, that unconventional engines have any chance to prosper. Such applications are rare and most of them are military, although marine engineers at various levels from the outboard motor to the giant ocean-going diesel seem to be surprisingly open-minded.

The history of the internal combustion engine is therefore replete with lost causes. Some of them reached the bench, some got no further than a patent application. All are worthy of consideration by the engine designer of today, particularly if he be designing for tomorrow as he should.

Unfortunately it is one of the perversities of historians that they will more aggressively dispute matters of fact than they will matters of opinion. It appeals to them more to be categorical about who was first with rotary valves, say, or opposed pistons than to pontificate on the desirability of such features.

It was by no means in this spirit, fortunately, that the Institution of Mechanical Engineers promoted a half-day discussion in December 1971 of *Some unusual engines old and new*. Sponsored by the Automobile Division of the Institution, this meeting was intended to be interesting, informative, and with any luck perhaps a little inspiring.

The proceedings took the form of a scrutiny (in the form of a series of addresses and subsequent discussion) of past, present and potential future attempts to break away from convention in engine design. It was to be an appraisal of features that failed and of the viability of ideas yet to be proven.

There are many 'characters' in the Automobile Division, especially among the engines men, who dearly love such opportunities to be controversial, to chide or poke fun at their most respected and liked fellow members, and if possible to trail the occasional red herring across any promising path. With the elements of present and future included in the discussion, lively interchanges of this sort were virtually guaranteed; whereas, had the original idea of considering merely *engines that might have been* remained the object of the exercise, the occasion might well have been a good deal more dull. Indeed, one of the contributors pointed out in negotiations preceding the meeting that the bulk of his work was on *engines that might well be;* and it seemed to me essential that this aspect of the half-day's study should be given its full weight in the book that the Institution asked me to base on the meeting. The programme for the meeting served ample notice that the object was to profit from the past for the benefit of the future, rather than to confirm parched historical facts.

It was essentially a programme of practical interest, with no more than a proper measure of old times for old times' sake. Thus the first session comprised two addresses on rotary valves, the first by M E Cross (son of a highly respected former Chairman of the Automobile Division, Roland Cross) dealing with the rotary valve engines that had been designed and developed by his father and which were no longer considered generally sufficiently attractive to pursue farther; and in this respect they were in marked contrast to the rotary valve engines designed, made and still under development by the second speaker, F M Aspin. Then there was C D Brewer talking about the stratified-charge engine, still essentially a laboratory experiment but nevertheless enjoying intense interest on the part of the environmentalists, and Professor F J Wallace discussing the chequered history of the free-piston engine. The next two speakers really needed to be treated as one, so close were their associations in Coventry Climax and Jaguar: W Hassan took as his subject the 16-cylinder racing engine produced but never put into contention by the former company, while H Mundy discussed the principles underlying his inchoate design for a 12-cylinder racing engine from which Jaguar cars formally and understandably wish to be dissociated. It was difficult to decide which of these two might be 'dum and which 'dee, but there was no doubt of their joint and several abilities to be entertaining, instructive, and more than a little provocative. It was rather interesting that they should be followed by another man with a distinguished record in motor cycle and car racing engines. L Kuzmicki, now with Chrysler and talking of the multi-bank tank engine that this firm put together during the war. As an example of unusual format, this could only be rivalled by the last topic of the programme, the Bristol wobble-plate engine whose intriguing principles were related by G W White, emerging from retirement for the occasion.

It was a proper mixture of the backward glance and the forward look, of dutiful reference to the awful warning and wistful deference to the needs of the future. Only in this way could it be as inspirational as was intended, for the history of the internal combustion engine is the story of intense application of scientific principle, of frequent indulgence of personal whim, and of occasional intervention of grievous misfortune. To the person with an orderly mind it may appear the most ill-assorted hash of repeated errors compounded by incompetence and intolerance; but it is nonetheless a romantic story of achievement and effort, of idealism and inventiveness. It is a story of more than metal and oil: human foibles, personal relationships, national pride and commercial greed all play their part. In the past these ingredients were more thoroughly mixed than in today's era of convergent technologies and carbon-copy design. A look back reveals no dearth of experiment in kaleidoscopic variety: there were several types that were manifestly good, but there was never any question of the possibility of some other type proving better, nor any shortage of freethinkers brave enough to try it.

There remain from those days many bright ideas to be pursued, many theories of construction that would be valid with our present materials even though when originated they were undermined by contemporary metallurgical or petrochemical shortcomings. Some that were successful enough in their times could perhaps profitably be revived; others that were abject failures could with equal profit be studied, for in them may lurk the germ of an idea that has yet to be properly developed.

Will it happen? One might be forgiven for doubting it, in these days when tooling costs a small fortune and a mistaken bit of missionary zeal may cost a large one. Though our attitudes to the internal combustion engine are changing as its place in the world it has created is changing, it is arguable that there is more hope now than ever before for some unusual engine to emerge that will satisfy the wants that the usual engine has notably failed to recognise. There is an enormous fund of experience and imagination tied up in the annals of the internal combustion engine; and as long as we continue to use it in some form or another there will be ideas to be harvested and lessons to be inferred from the records that have already been logged. Perhaps advocacy will constitute the sternest barrier, for things are nowadays designed by committees, and such bodies are not easily persuaded out of their clearly defined grooves.

This then is a book not only about old unusual engines but also about new ones, not only of engines that might have been but also of engines that might well be. It takes into account the pattern of history which is recognisable as dividing engines into classes that are unusual in some particular respect. There are those that are unusual in format, others unusual in construction; many are defiant of convention in their valve gear, some resolute in their divergence from normal means of translating power; there is an interesting class of engines whose designers have despaired of nature's inefficiency in charging and scavenging, relying instead on original and unnatural means of accomplishing these ends. There are a few engines that might be fitted into more than one of these categories, the Napier Deltic providing a good example – as it does of an unusual engine that is far from being a commercial failure. Several other engines that will come within our review prove quite conclusively that being unusual is no bar to being used.

1. Engines of Unusual Respiration in Charging and Scavenging

If the good Lord had ever intended man to fly at high altitudes, He would have made the atmosphere of constant density. However, man has always been one to fly in the face of providence, and of the many by-products of this defiance we must remark upon the supercharger. It is merely a mechanical contrivance to compensate for the deficiencies of nature.

Consider the problems peculiar to high-altitude aviation and it is easy to see why the supercharger was adopted for aero engines at an early stage in the petrol engine's career, long before it enjoyed much custom in other applications. Even so, there have been such a lot of most unusual supercharger arrangements contrived, not only for aero engines but also for marine and road traction, that it might have seemed evidence of negligence that the programme for discussion at the Institution's London headquarters in December 1971 included no engines relying on anything other than natural atmospheric induction. In fact this was no fault of those who planned the meeting, for one of the authors – Professor Wallace – was known to be doing some fascinating work on the differential compound supercharged engine, and was originally invited to participate with an address concerning it. At that time, however, his work on this engine had reached the stage where Professor Wallace felt that a comprehensive research paper suitable for an ordinary meeting might soon be desirable; and he therefore opted to discuss instead the free-piston engine – an apparatus which could perhaps be deemed supercharged, and one so extremely unusual that the organisers were quick to welcome the suggestion.

Three years earlier, Professor Wallace had discussed progress on the differential compound at a symposium run by the Institution at Nottingham under the title *Transport engines of exceptionally high specific output*. The engines under discussion there ranged from the H16 3-litre BRM to tank engines, marine engines and railway traction engines; but the most remarkable feature of the symposium was that, of the eight engines discussed, four were supercharged two-stroke diesels and three of those were of the opposed-piston variety. At the time it was remarked that another paper (delivered a year earlier still) demonstrated that if you want something really frantic in the way of power outputs this is the kind of engine with which to achieve them.

This may prompt the more retrospective readers to recall the type 451 Fiat, built and developed in 1925 and 1926 for the then current 1½-litres Grand Prix motor racing formula but never raced. Like so many other opposed-piston two-strokes, it was based on principles expounded in the Ochelhauser gas engine of the 19th century, with its twelve pistons opposing each other in six cylinders that were ported according to the demands of the two-stroke cycle, scavenge blowing being provided by a Roots compressor delivering air at 0·37 atmospheres (5 lb/in^2) above ambient. This air was delivered from the blower through the carburettor (a mistaken practice to which the Italians and Germans were too long addicted) to the piston-controlled inlet ports. The engine developed more than 150 bhp, a power that no rival manufacturer could summon from 1½ litres at that time; but it suffered from severe overheating in and around the exhaust pistons, causing the Fiat engineers to contract severe impairment of the enthusiasm. It is debatable whether, if such an engine were to be designed today in the light of current knowledge, it might be expected to prove acceptably efficient and reliable. In the case of diesels there can hardly remain any such doubt.

It was in 1963 that Professor Wallace first developed the concept of the differential compound engine, in a theoretical paper based on earlier studies of the potential performance of opposed-piston two-stroke diesel engines subjected to very high boost pressures. These studies have indicated that an engine of this type, which gave 120 bhp when naturally aspirated, might after modification with variable-compression-ratio pistons achieve ratings of 320 bhp at a boost pressure ratio of 3, rising to as much as 420 bhp at a pressure ratio of 5. Such power in such modest bulk (it represented 29 to 37 bhp/ft^3) represented a worthwhile advance even over the best results achieved in the high-output four-stroke diesels introduced by such American firms as Continental and Caterpillar in the 1960s; and it was judged worthwhile to embark on the appropriate research at Bath University of Technology (Bristol), using as a basis the well known Rootes TS3 compression-ignition engine.

The differential compound engine is one in which three basic constituents (engine, compressor and turbine) are connected through differential gearing. The exhaust gases from the engine are fed to a turbine, the output shaft of which is geared to the output shaft of the engine by way of the spider carrying the planetary pinions of a differential gear in which the annulus is on the engine output shaft and the sun wheel drives the superchargers.

Courtesy: Fiat Centro Storico.
Worthington-Williams collection.

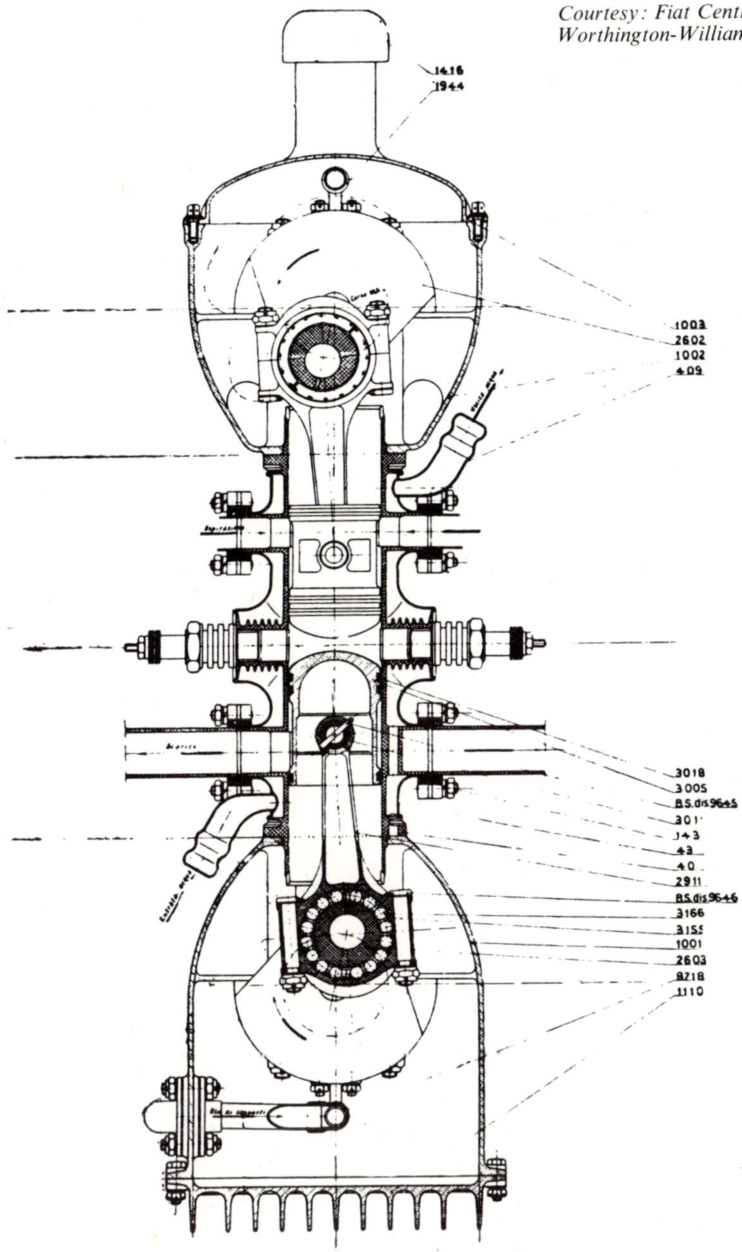

The 1925 Tipo 451 Fiat illustrated here was built for the $1\frac{1}{2}$ litre GP formula, but never raced. An opposed-piston six-cylinder two-stroke it employed the nineteenth-century Ochelhauser principle (in common with the Junkers Jumo, the Napier Deltic, and the Rootes commercial vehicle unit now being considered for licence production by Armstrong Whitworth).

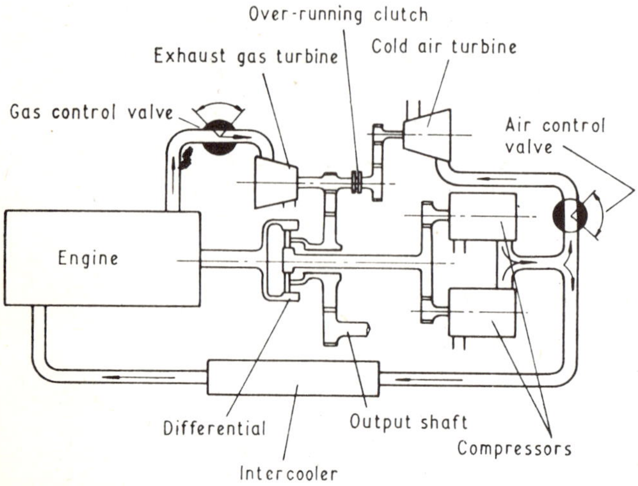

The Differential Compound Engine developed by Professor Wallace represents the successful transition from theory to practice which so often eludes engine designers. Its performance continues to justify fully the outputs predicted in the Professor's theoretical papers, which were based on earlier studies. These had disclosed that opposed-piston compression-ignition engines operating on the two-cycle principle and under conditions of very high boost would return outputs substantially higher than the best then recorded by the latest four-stroke high-output diesels.

Using a Rootes TS3 diesel engine coupled to a compressor and turbine (which form, with the engine, the aggregate of the compound principle) by means of differential gearing, it was found that by incorporating an auxiliary turbine, very high torques at low output shaft speeds were achieved. This, together with the high potential outputs possible, utilising the differential compound principle, make the system worthy of detailed consideration for road and rail application.

Initial tests broadly confirmed the predictions of torque/speed and power/speed relationships, which were expected to be appropriate to traction-type prime movers. The idea was to eliminate the need for multi-ratio gearboxes or their hydrostatic or hydrokinetic equivalents; and the feasibility of this was publicly demonstrated in 1964 by Perkins in a truck powered by what they simply called the differential diesel. The arrangement used by Professor Wallace secured particularly high torque at low output shaft speeds (the Perkins proved to need a two-speed gearbox for fully acceptable performance) by the incorporation of an auxiliary cold-air turbine which was fed by excess air from the superchargers, and which in turn fed its output torque into the output shaft of the exhaust gas turbine by way of an overrunning clutch.

Impressive though the interim test results were, in proving that the basic concept was substantially realised, the actual efficiencies, powers, and torques, fell short of predictions – particularly at high output shaft speeds and boost pressure ratios. Much of this was due to severe parasitic losses, many of which would be eliminated in production units; and it was forecast that the incorporation of variable geometry for the power turbine, not to mention interburning between the compressor and auxiliary turbine, would materially improve the performance of the engine.

There are some interesting echoes here of aviation practice, especially that of the distinguished and brilliant engineers who once made the name of Napier one to treat with the utmost respect. At the Nottingham symposium the underlying, almost subliminal, message seemed to be that the turbocharger was a good thing but not good enough, whereas if it be compounded by some mechanical nexus between impeller and crankshaft very good results could be achieved at all times, together with admirable response and the highest possible thermal efficiency. Every one of the blown two-stroke quartet employed such a system, as did some years earlier that exceptional engine the Napier Nomad which was also a blown two-stroke diesel. The principle can be applied to petrol engines as well, as the Wright Turbo Compound would serve to remind us.

The history of the idea goes back somewhat further to the 1920s, when the Engine Division of the Bristol Aeroplane Company was experimenting with the exhaust turbocharger with much more enthusiasm than had been displayed by any other aero-engine manufacturers. Of course there had been exhaust turbochargers on aero engines before then, most of them significant failures due to the metallurgical shortcomings of the times; but by 1924 Bristol were actually flying a Jupiter radial with turbocharger, operating it successfully and with great promise at altitudes above 30 000 ft. For various good reason that here do not concern us, work was then suspended until

1937, by which time they were well advanced with their sleeve valves; and it occurred to Bristol that, since the sleeve valve was the more suited to operation at high back pressure, the objections sometimes raised against exhaust turbines were of less moment. Thus encouraged, they began investigation of a turbocharged version of the sleeve valve engine, the idea being that boost could be progressively increased with back pressure. Contemplation of this relationship led to the realisation that ultimately a state of affairs might be reached where the exhaust turbine was as powerful as the engine: and at that point it becomes logical and possible to transpose the airscrew and the compressor element of the turbo engine. Thus the engine and compressor would constitute a gas generator unit, while the turbine would translate the gas energy into power for driving the airscrew. Already one can see here vivid analogies to the free-piston engine, which again was virtually a gas generator feeding a power turbine. The only trouble with this design philosophy is that if you pursue it too rigorously you are in grave danger of losing your pistons altogether and inventing the gas turbine, which might not be what you first intended.

That is what happened to Bristol: by 1939 they had already made plans for getting into the gas turbine business. However, the idea of the compounded piston engine had been germinated for a long time, and it spread to other manufacturers who adopted it with great enthusiasm.

Most outstanding of these was Wright, who applied the principle to their R3350 Cyclone radial. At the time, this engine gave up to 2800 hp at take-off and weighed 3029 lb. The modifications adopted by Wright increased its performance considerably, while keeping the engine dimensionally so similar to the standard version (the diameter remained unchanged, the overall length grew by only a few inches) that no special installational difficulties were presented. To cope with the greater power the crankshaft was stiffened and cooling fin area increased by forty per cent. The next step was to endow it with a two-speed gear-driven supercharger, a slightly unexpected stratagem which made boost and back pressure to some extent independent. The exhaust pipes from each of the eighteen cylinders were then led in clusters of six to three blow-down exhaust turbines which were linked to the crankshaft by fluid couplings. Between them these three turbines returned about 550 hp to the engine, sometimes more. The performance was so improved that the take-off rating was raised to 3500 hp, or even 3700 hp in a special version for the US Navy. Perhaps more important was that the specific fuel consumption during cruising was only 0·38 lb/hp h.

This meant that an aircraft in which the Cyclone was replaced by the Turbo Compound could fly 20 per cent farther for the same fuel (the Lockheed

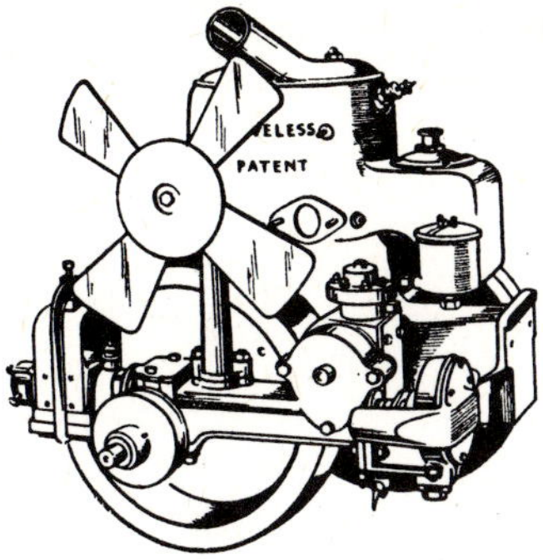

FIG. 1

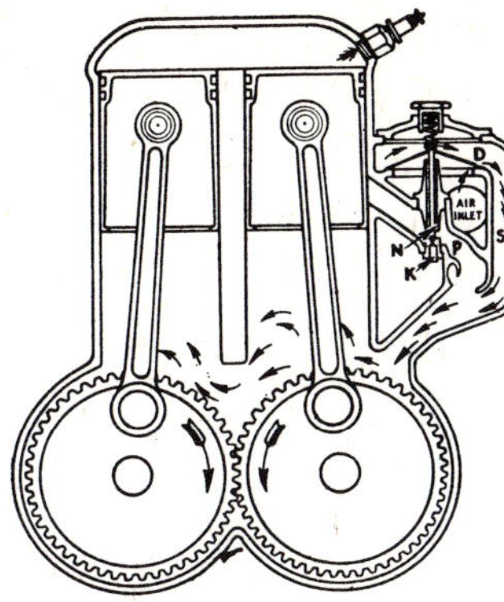

FIG. 2

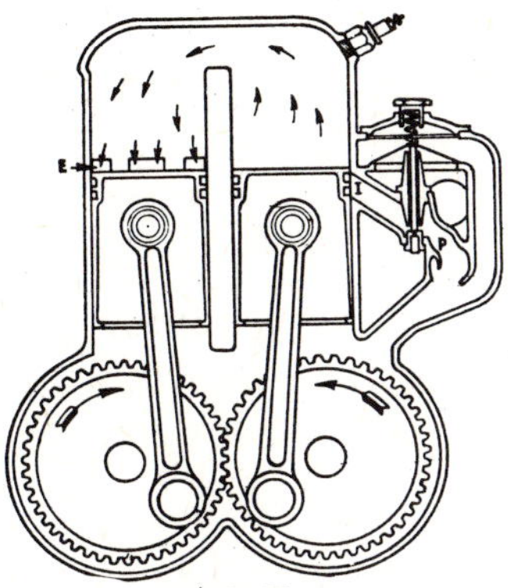

FIG. 3

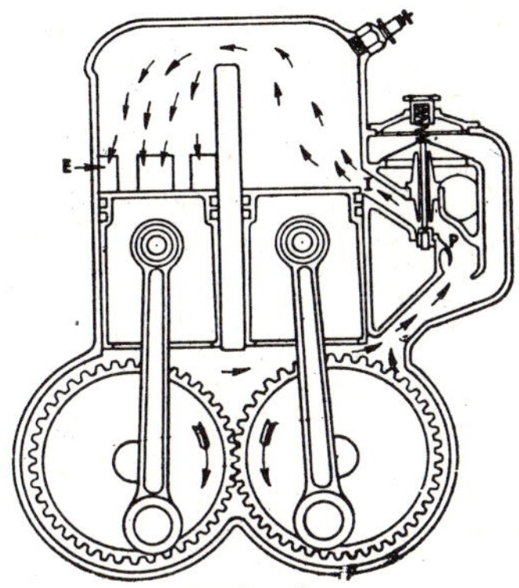

FIG. 4

Produced in 15·8 hp, 19·9 hp and 25 hp form, the Valveless two-stroke-engined car produced between 1907 and 1913 marked the first essay into motor manufacture of the David Brown Group of Companies. It featured crankcase compression, geared twin cranks, and twin pistons with a common combustion chamber.

Courtesy: David Brown Gear Industries Ltd.

Gyrocar: The twin flywheels of the Valveless engine can be clearly seen in this photograph of the example recently restored by the David Brown apprentices. Note also the linked twin sparking plugs. The large fan and header tank, whilst indicating possible cooling problems, are probably not standard since the restored car was brought back from South Africa and was probably a 'Colonial' model.

Courtesy: David Brown Gear Industries Ltd.

(*see right*)
Utilising a two-stroke principle invented by J Day in 1891, the 1910 NEC (New Engine Company) vee-four aero engine is worthy of note in that it was the first aero-engine to employ supercharging. This sectioned engine shows the Roots-type blower which the NEC's two-stroke system necessitated: this positive-displacement blower was a substitute for crankcase compression.

Courtesy: Science Museum, London.

Super Constellation had its range extended up to as much as 5840 miles), and so the success of the venture was assured. On the other hand, a far more thorough (not to say brilliant) exploitation of compounding principles by Napier enjoyed no success at all: their Nomad engine, conceived at a time when propellors were expected to be the normal means of air liner propulsion, did not mature until the aviation world had gone on to jets. Like so many unusual engines, it arrived too late; and to be truthful it is arguable that it never really matured anyway, since it was by all accounts a pig to start.

Looking at its specification, this hardly seems surprising. The Nomad was a 12-cylinder horizontally-opposed liquid-cooled two-stroke compression-ignition engine compounded with an exhaust gas turbine, both of these power units driving the common single propellor shaft through reduction gears. Even the basic construction was satisfying: the crankcase was a two-piece structure of magnesium alloy castings, the two cylinder blocks were of aluminium alloy with dry liners fitted in each cylinder bore. Each of the cylinders had its own aluminium alloy head, elegant and simple in shape because the eight inlet and the three exhaust ports of each cylinder were of course in the walls.

As appropriate to a two-stroke, the ratio of bore to stroke was unfashionably low, resulting from measurements of 6 and 7·375 inches respectively. This yielded a displacement of 2505 in^3 (41·1 litres), making a fairly big engine that weighed 3580 lb. Beneath and behind the crankcase was the turbine department, where a three-stage axial-flow turbine rotor was mounted on a shaft which drove through a variable ratio Baier fluid coupling and gearing which connected it with the propellor shaft – which in turn connected through reduction gears to the crankshaft. Coupled to the turbine shaft was the compressor, a twelve-stage axial-flow affair delivering air to the cylinders at very high pressure (8·25 atmosphere) and in enormous quantities (13 lb/s at maximum speed).

Many an engine of much less complication has been debased by some want of efficiency in one or more of its component elements. It is a tribute to the purposeful design of the Nomad that, with so many constituent sections that could have let it down, it was in fact of extraordinary efficiency. The whole operating cycle was calculated to extract every possible quantum of energy: nothing was allowed to be wasted at any stage. After combustion had been initiated by the injection of diesel fuel in the cylinders, the initial expansion of the charge would deliver power through the pistons to the crankshaft. As soon as the exhaust ports were uncovered expansion would continue through the exhaust manifold to the turbine, where the gases and residual hot air produced by combustion would liberate more power for

Courtesy: Curtiss-Wright Corp.

An elaboration of the Wright R 3350, the EA Series Curtiss-Wright Turbo Compound shown here was virtually identical in principle to the later Napier Nomad.

Courtesy: Curtiss-Wright Corp.

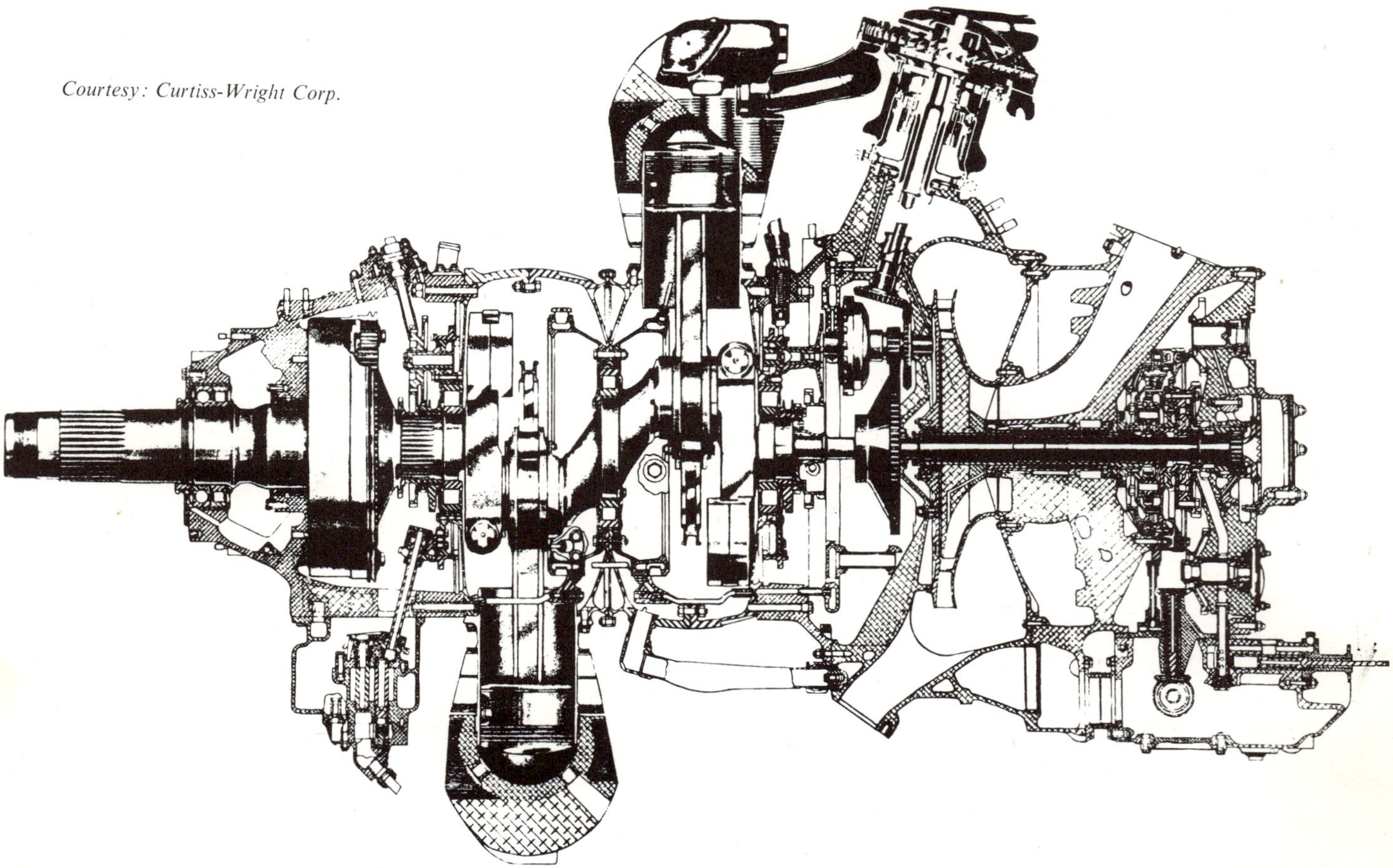

Wright R3350/32W Turbo Compound radial. Developed from an earlier Cyclone, the R 3350 with 18 cylinders, it had three blow-down turbines for exhaust gas power recovery set at 120 degrees to one another, and incorporated not only a reverse low flow torquemeter and two speed impeller drive, but automatic spark advance and impeller fuel and water injection systems for take-off or emergency boost as well.

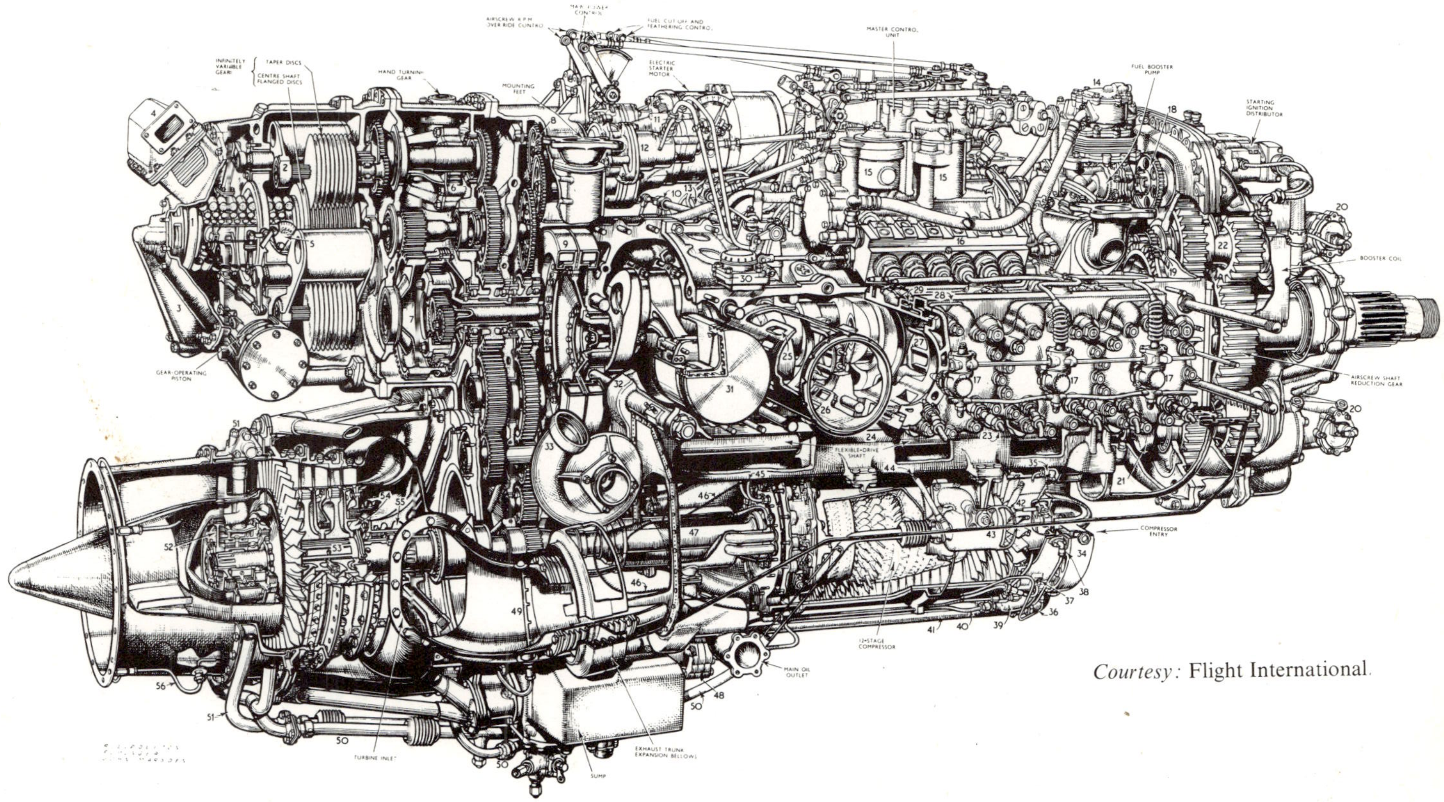

Courtesy: Flight International.

Conceived in good time, born too late. The Napier Nomad with its opposed-piston layout, liquid cooling and apparent complexity employed a diesel two-stroke system compounded with an exhaust gas turbine. It can be contrasted with the compact and successful Wright R 3350.

Unorthodoxy triumphant: Developing 12 hp at only 1000 rev/min, this 1928 Trojan differed little from the prototype designed by Leslie Hounsfield in 1910 and was still current in 1939. Post-war Trojans reverted to the even older Ricardo Dolphin principle.

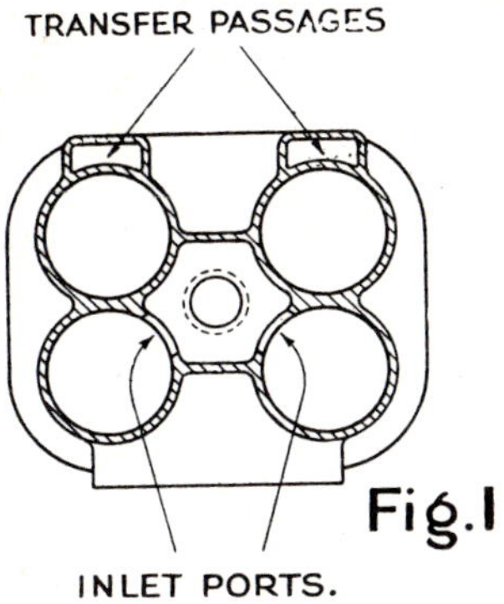

Fig.1

SECTION THROUGH AA.

Fig. 2. I Carburetted air entering vacuous crankcase

IV Compression complete, spark explodes the charge

Fig. 3 II. Compressed explosive mixture in crankcase about to enter cylinder through transfer port.

V Explosion complete, burnt gases escaping through exhaust port. The new charge is about to enter and take the place of the burnt gases.

Fig. 4 III Transfer from crankcase to cylinders complete, and compression in cylinders about to commence.

Worthington-Williams Collection.

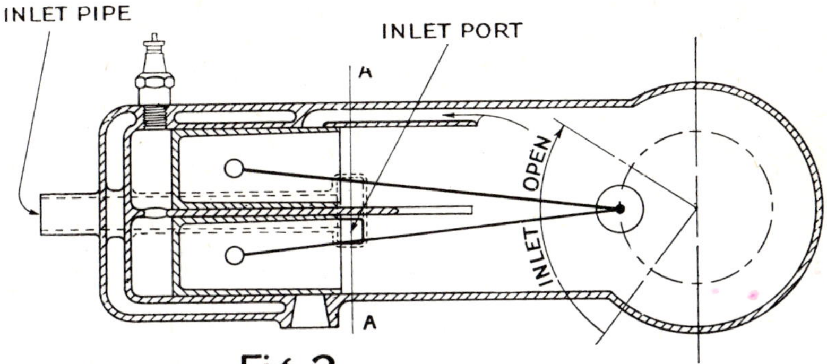

Fig.2

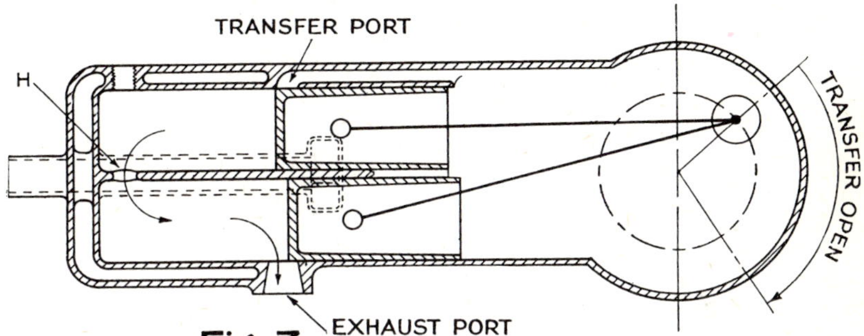

Fig.3

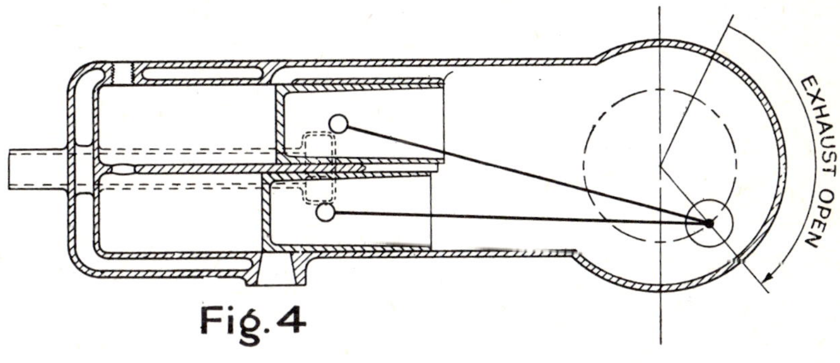

Fig.4

transmission through the hydraulic coupling to the propellor shaft. The total power from crankshaft and turbine together was considerable, and with water injection the take-off rating was 3476 hp at 2050 rev/min. But there was more to come: there was still a little energy left in the exhaust gases even after negotiating the turbine, and this was squirted out as a jet at the back to provide a further 250 lb of thrust, making a total equivalent horsepower of 3570. This was equivalent to a maximum bmep of 205 lb/in^2, a very high figure for a two-stroke. The other specific performance factors were no less impressive: the engine weighed virtually one pound per horsepower and developed 10·5 hp for every square inch of piston area – which provides a revealing comparison with the 6·58 hp/in^2 of the Wright Turbo Compound. At maximum continuous rating the Nomad developed 2248 equivalent hp; but looming overall was an incomparably mean specific fuel consumption. The engine had after all been conceived as the propulsive unit for a really long range aircraft, intended to realise the most outstanding economy. Napier claimed 0·33 lb/hp h, although Air Vice Marshal Banks has hinted to me that the engine never quite achieved that.

It was nevertheless a most satisfying performance. Napier were not to be satisfied, however, for they argued that it should be possible to do something with the unburnt air in the exhaust system. A diesel can only burn 70 per cent of the air it breathes: they therefore inserted an afterburner nozzle in the exhaust manifold, injecting extra fuel to burn the remaining oxygen and thus allow the turbine to make a much greater contribution to the total output. This and an intercooler between the compressor and the cylinders added a mere 170 lb to the weight of the engine; but the return for this slight investment was a fantastic profit of no less than 530 hp. On this basis the specific weight fell to 0·83 lb/hp – and who would have thought that any diesel would have proved to be relatively lighter than the majority of spark-ignition engines gulping relatively larger quantities of the best quality petrol?

In a rather simpler form, the idea of the exhaust turbine positively linked to the crankshaft has been revived comparatively recently by Mitsubishi in their 10ZF tank engine. This is an oddity in several respects: although it is a high-output two-stroke diesel, it is air-cooled; although as diesels go it is a fairly high-speed affair, it has ten cylinders; and although its specific output of 858 hp from 21·5 litres is high in its class, it is extremely compact.

Apart from the air cooling, which is supplemented by partial oil cooling of the cylinder head and forced oil cooling of the piston, the principle feature of this V10 engine is that there are two exhaust turbochargers at the front end of the engine, driven mechanically through a torque-limited clutch by the crankshaft. The object of this was to the eliminate the positive displace-

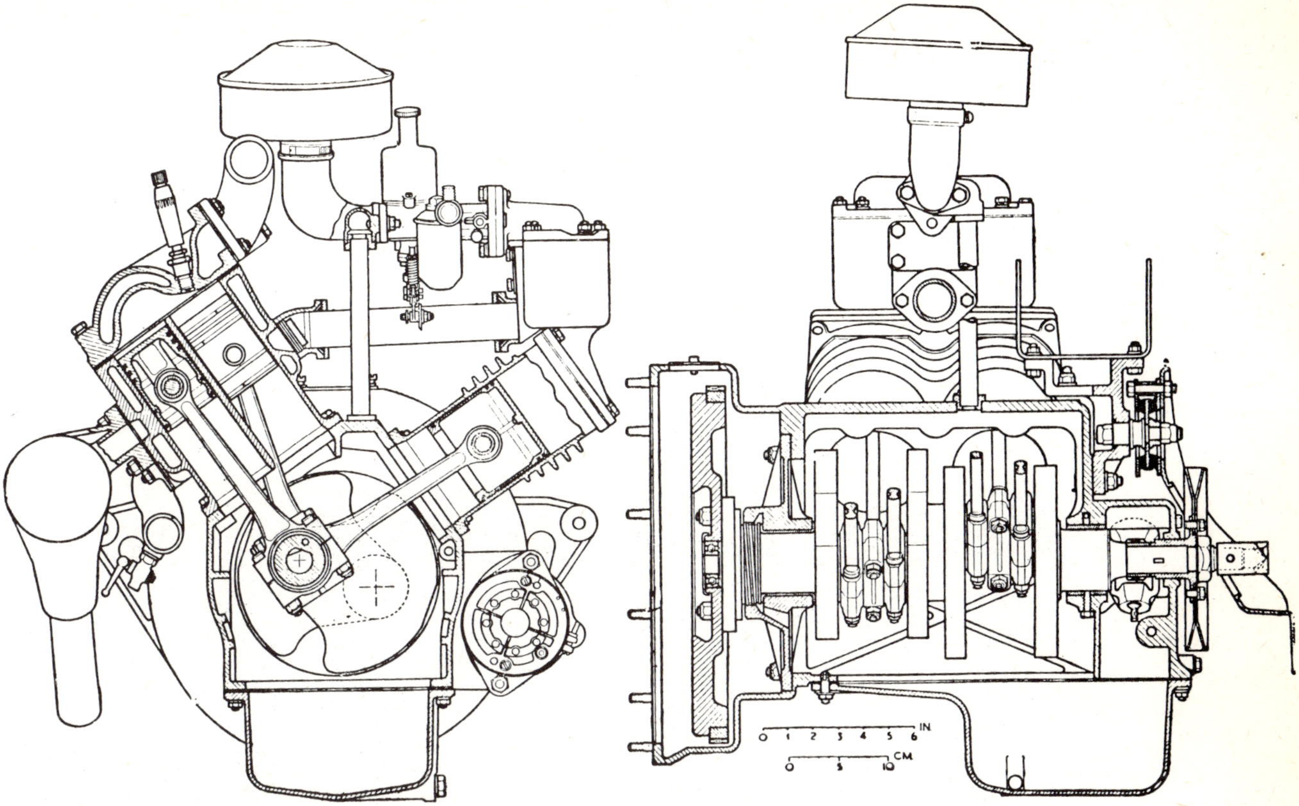

It must have been a source of some satisfaction to Ricardo that two years before his knighthood in 1948 he was to see his original 1906 Dolphin design vindicated by the introduction of the Trojan 65 light commercial unit.

Incorporating the paired cylinder two-stroke principle employed on the pre-war models (it employed a number of parts common to the earlier type) it eschewed crankcase compression in favour of Ricardo's pumping cylinder, arranged (as in the Dolphin) to give a Vee appearance.

It remained in production for over a decade, although latterly the two-stroke was available only on export models, the home market having to content itself with Perkins diesel versions.

Courtesy: E Wright and Trojan OC.

ment type of scavenger blowers which are necessary for conventional high-speed two-stroke engines, and to keep the supercharging system as compact as possible. An interesting feature of the exhaust turbochargers is that, whilst the pump element is centrifugal as is usual, the turbine is an axial-flow type. The torque-limiting clutch is there simply to protect the driving system from overloads, not to modulate the performance.

Mitsubishi gave some interesting comparisons of the mechanically driven turbocharger system and the more conventional arrangement of Roots blower in series with the turbocharger system. With the Roots blower in series with turbocharger in the higher speed range, the rotating speed of the turbocharger was increased as the engine load was increased, resulting in higher air pressure at the delivery from the impeller and hence a higher pressure at the inlet of the Roots blower. It follows that at a higher engine speed the power necessary for driving the Roots blower would increase as the engine load increased. With a mechanically driven turbocharger system, on the other hand, the exhaust gas temperature and pressure are increased as the engine load increases. Thus the turbine draws more power from exhaust gas, with the result that the power required to drive the blower impeller is reduced. It is worth pointing out that if these two systems are designed to give the same scavenging air pressure at full load, the mechanically driven turbocharger system will be required to consume more power at the partial load condition compared with the other system.

These arrangements in the Mitsubishi engine are not without their shortcomings. Determining the gear ratio between crankshaft and turbine shaft brought its headaches: raising the ratio to give a higher turbine speed might well produce more scavenging air, a lower exhaust temperature, and the opportunity to increase maximum power output by increasing the quantity of fuel injected, limited only by peak exhaust gas temperature. However, the maximum explosion pressure in the cylinder would be raised and would give a poor specific fuel consumption at part load owing to the increase in power needed to drive the blower. Even the format of the engine brought its problems: Mitsubishi soon discovered that the exhaust manifold could not be entirely satisfactory, in that they were virtually forced to use simple five-cylinder manifolds chosen for compactness rather than for anything else. No explanation was ever given for the choice of the number of cylinders, though it is known that all two-stroke V-engines with even firing and with less than twelve cylinders have some degree of imbalance with regard to either primary or secondary moments if not both. Indeed it was a Japanese, Nakanishi of Tokyo Imperial University, who clarified this as long ago as 1932.

Looking back over the years, we see more than a little evidence to suggest

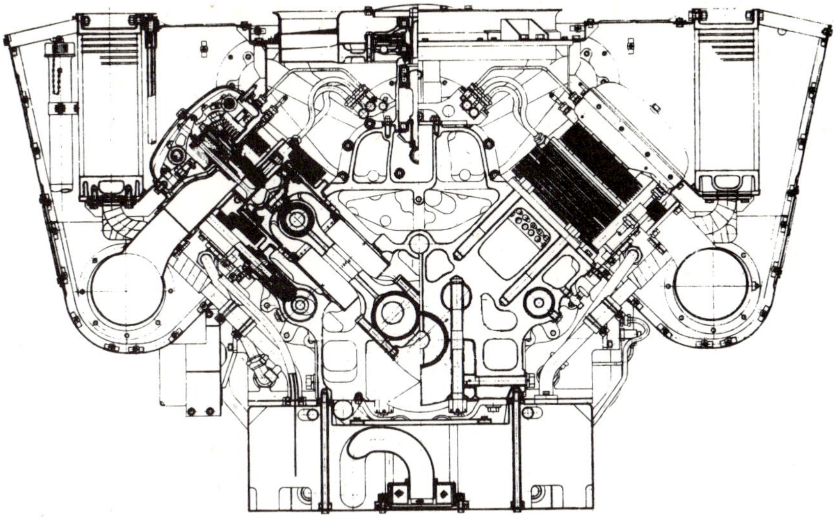

Developed for military tank use following exhaustive preliminary field work with single and four cylinder prototypes, the Mitsubishi 10 ZF 10-cylinder Vee formation air-cooled two-stroke diesel is particularly noteworthy in that despite its high output (858 hp from 21·5 litres at 2200 rpm) its designers have overcome the cooling problems which in the past tended to limit output from this type of engine to a maximum of 100 hp.

Cooling is catered for by two sets of fans located horizontally in the vee cavity between the cylinder banks, and in addition four oil-coolers are installed along both sides of the engine (two each side) one of which is for the transmission oil. Cylinders are finned and positively cooled by lubricating oil which circulates the machined passages between the valve ports and fuel injector inlet. Performance is enhanced by the employment of a uniflow scavenging system and direct fuel injection, exhaust turbochargers of the centrifugal blower and axial turbine type, and four mechanically-operated exhaust valves per cylinder.

that engine designers have been perhaps too preoccupied with controlling the behaviour of the air on its way to and from the cylinders while leaving it at the crucial intermediary stage to revert to its Shakespearian role as a chartered libertine. Nowadays, with environmentalists and every kind of political panic-monger screaming for low emissions and low fuel consumption, this approach is hardly good enough. Admittedly a good deal of work is being done on the turbocharged engine with a view to satisfying the requirements of these intrinsically conflicting parties, while the ordinary supercharger is almost certain to be shunned for all spark-ignition engines, on account of the relatively low thermal efficiency commonly associated with it. Nevertheless there are other and more simple ways of achieving similar ends, and one of the most interesting of these, if not yet the most practical, is the stratified charge engine. Ford have been notable in the development of this, and the paper by Charles D Brewer explained just why and how Ford had gone about it. The process of stratified-charge combustion had been described by Witzky as that of a fuel-injected unthrottled spark-ignited cycle that requires selected stratification of the fuel around the spark plug to provide a fuel mixture within the ignitable air/fuel range, to accomplish the ignition of the fuel charge at low power outputs when the overall air/fuel ratio may be of the order of 100:1, and still avoid excessively over-rich mixtures at the spark plug at full power. In other words we have something that is a cross between a diesel and an ordinary petrol engine, burning mixtures that the latter would not normally tolerate and constructed with an economy that the diesel could not long endure. Ford assumed that to be commercially viable the stratified charge system would have to provide fuel economy characteristics equal to or near theoretical limits, smoke-free power equivalent to carburetted engines when operated on ordinary petrol, together with identical speed and torque characteristics, not to mention levels of harshness moderate enough to be suitable for passenger cars. Other important points included adquate cold starting characteristics without the use of larger starters or batteries or other cold starting aids such as a choke, while a high degree of commonality of engine parts with ordinary carburetted engines was also desirable; finally the emission characteristics had to be sufficiently low to meet anticipated standards.

In the Ford stratified charge engine all these objectives were substantially achieved, by the use of an open combustion chamber in the piston crown in conjunction with an inlet port giving plenty of swirl. The configuration is very much like that of a direct-injection compression-ignition engine with the addition of a spark plug to initiate combustion. So much use was made of existing manufacturing equipment and engine components that Ford were

An inability to idle at low engine speeds was a common fault in the nineteenth and early twentieth centuries particularly associated with the two-cycle principle. When, as a young undergraduate, Harry (later Sir Harry) Ricardo designed his two-stroke Dolphin engine (employing Dugald Clerk's theories) he overcame by means of a stratification 'bulb' what other manufacturers could only 'cure' by the untidy engineering expedient of fitting large flywheels. A slow 'tick-over' endeared the engine to fishermen using 'drift' nets, and some 50 were sold to the Shoreham, Sussex, fishing fleet for marine use.

Worthington-Williams Collection.

'The piper calls the tune'. When Harry Ricardo persuaded the wealthier members of his family to purchase cars fitted with his Dolphin engine, the bespoke coachwork they specified dictated that a larger redesigned engine would have to be employed. The result was the 28 hp 'Vee-eight', with four working cylinders and four pumping cylinders, shown here being installed in Dolphin cars in the Shoreham premises of the Two-Stroke Engine Company, circa 1908. One of them successfully carried Ralph Ricardo (a cousin) and his bride on their honeymoon to Scotland and back, in June 1909.

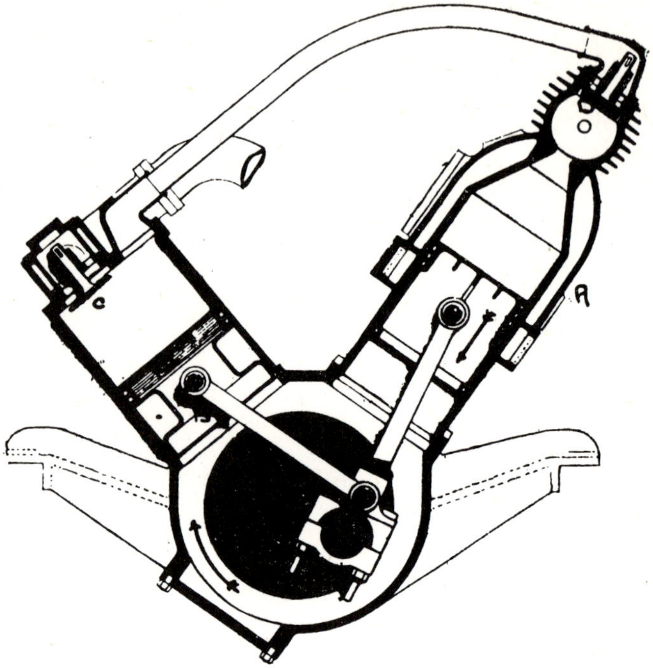

A section through the Dolphin engine showing both pumping and working cylinders. The stratification 'bulb' can also be seen, and the latter, when allied to an atmospheric inlet valve of the 'reed' type which allowed a *delayed* charge of air to be drawn into the combustion chamber by the vacuum action of the induction stroke, permitted smooth running almost without precedent in a two-cycle engine for the period without the necessity for cumbersome flywheels. This latter feature facilitated the adaptation of the engine to automobile use. A two cylinder 15 hp version was used by Harry Ricardo as personal transport until 1919, and his earliest experiments with anti-'knock' fuels at Walton-on-Thames (before the opening of the present-day Shoreham laboratory) were conducted using Dolphin engines of this type.

Worthington-Williams Collection.

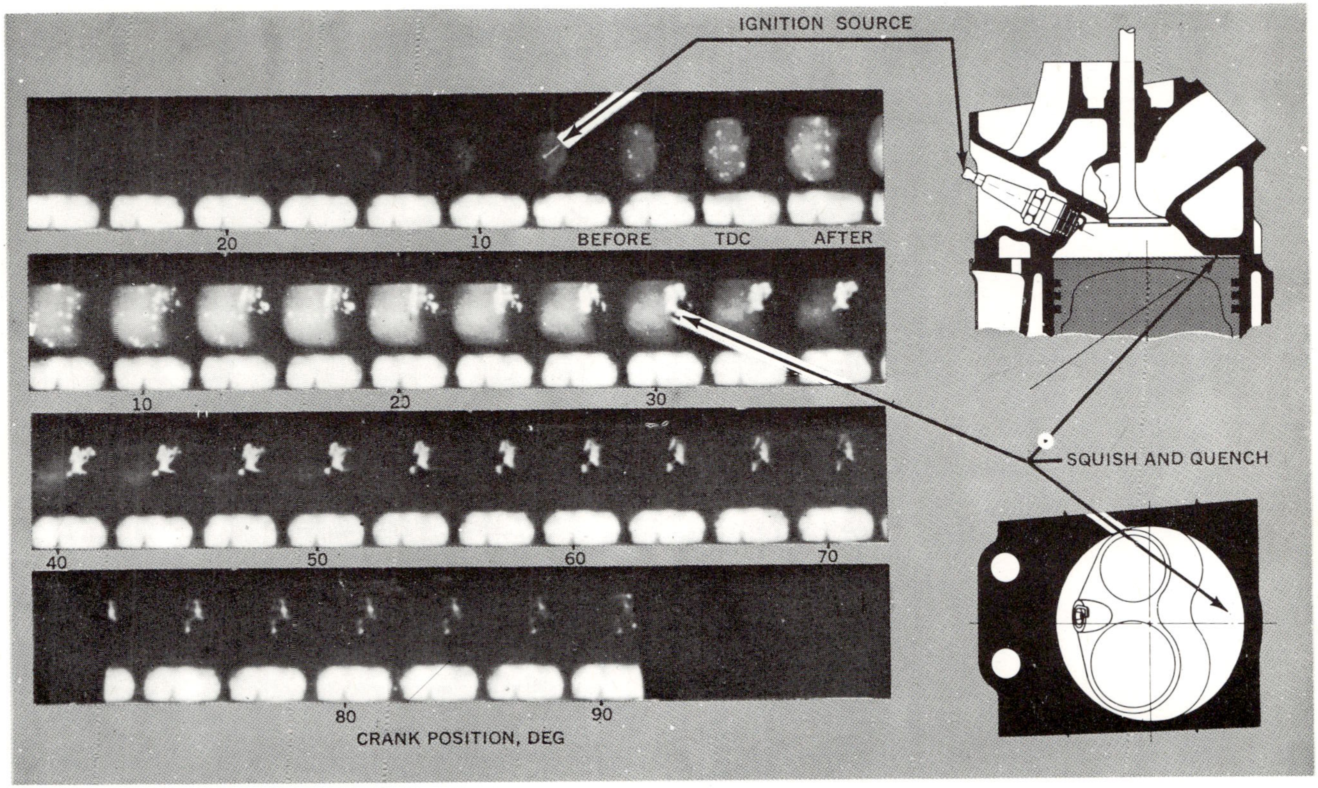

The Ford Witzky Stratified Charge engine, combining the advantages of both the Otto and Diesel cycles, is insensitive to fuel quality and can employ almost any hydrocarbon fuel. It eliminated to a large extent the problem of 'knock' in engines, which had engaged the attention of engineers for years. Ricardo designed a stratified charge engine in 1906 and his laboratory at Shoreham was set up in 1919 largely to study such problems. The principle has of course returned to prominence in view of current emission legislation, due to the engine's ability to utilise lean mixtures, thus reducing exhaust toxicity. See also p. 25.

Courtesy: Southwest Research Institute.

able to claim that many production engines could be converted to the FSC system by a change only in the cylinder head and piston, and of course in the provision of a drive for the fuel pump. Experimental conversions of V8 and in-line-4 engines were successfully installed in vehicles without structural alteration, and demonstrated overall savings in fuel consumption of as much as 31 per cent without loss in performance or increase of noise or vibration.

There was however an important caveat to enter. Success in the application of the stratified charge principle depends on the proper phasing of the fuel injection and the ignition timings. These change relatively with each other and in crank angle with every change in load and speed. Ford had developed fuel injection pumps and nozzles appropriate to the precise needs of the system, but the evolution of a computerised device that could properly exploit them promised to be extremely difficult. Another point noted by Brewer was that although the octane requirements of the engine could easily be met with commercial grades of petrol without the use of anti-knock lead compounds such as tetraethyl lead, the engine was sensitive to fuel volatility.

Brewer also pointed out that although the substantial saving in fuel consumption possible with the FSC engine had been amply demonstrated, motor manufacturers had shown little interest in it. Things are different now, but the inherent cleanliness of the engine from the point of view of atmospheric pollution did attract quite a lot of attention to the principle; and it might be prudent to examine this in a little more detail. In a conventional petrol engine, variations in load are met by variations in the quantity of charge while the ratio of air to fuel in that charge remains approximately constant. If this ratio could be varied and a full charge of air used with a variable amount of fuel, as in compression-ignition engines, the efficiency of fuel utilisation could be improved quite a lot, especially in part-load conditions. Alas, mixtures much weaker than about 15:1 ignite unreliably and at about 20:1 they generally do not ignite at all. In the stratified charge injection system, a rich mixture is kept in the immediate vicinity of the spark plug while a very lean mixture elsewhere in the combustion chamber is left to be consumed by the resulting flame.

The desired stratification can be achieved in a variety of ways, but premixing in a carburettor is clearly out of the question. There have been several other approaches as well as that of Ford, notably by Texaco, and development has also been pursued by Citroën, Renault, and the Institut Français de Petrole. Typically they contrived to impart a high-velocity swirl of the intake air through appropriate design of the air inlet port. During the compression stroke the fuel is injected at an angle back towards the swirl of the air, where it is entrained and carried towards the spark plug in

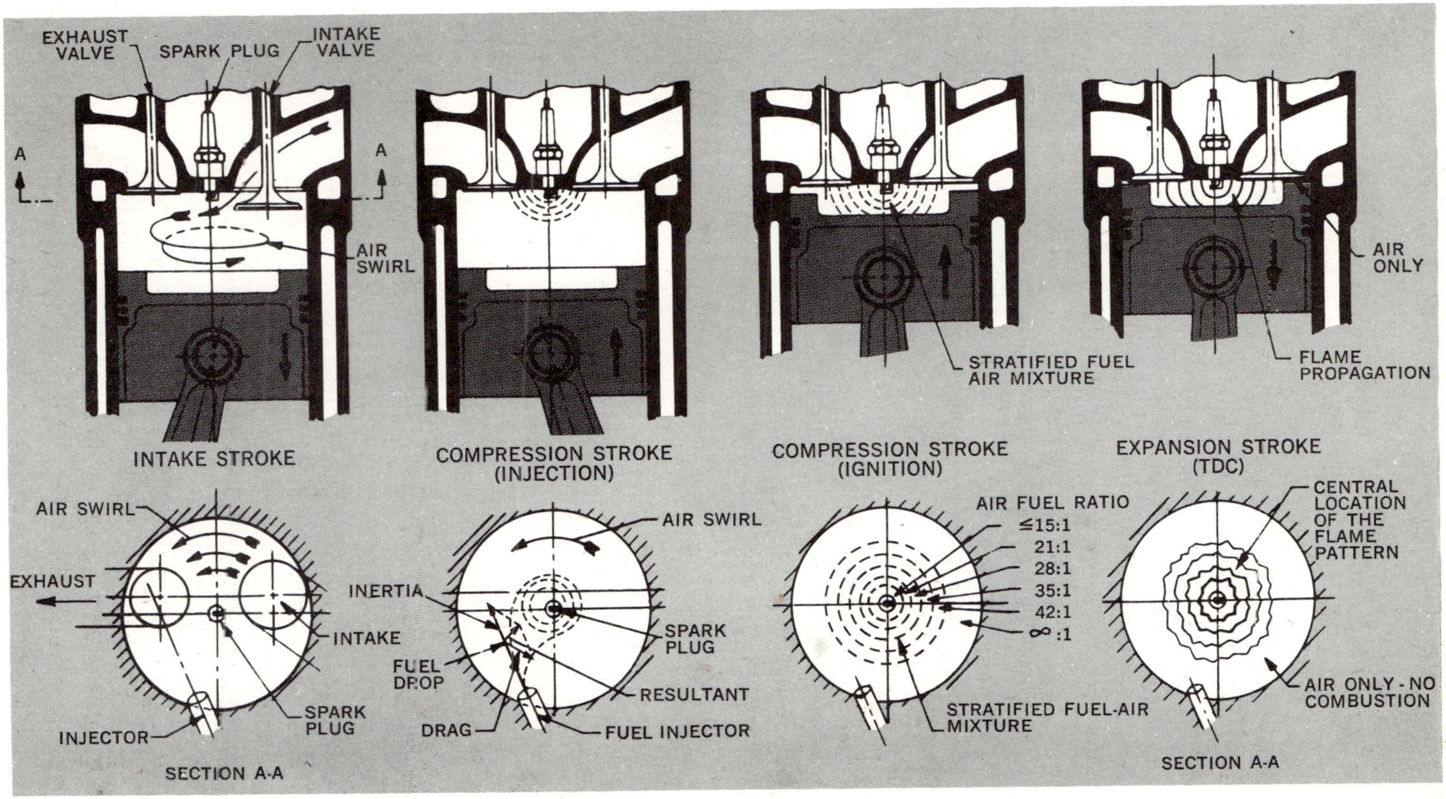
EXHAUST VALVE
SPARK PLUG
INTAKE VALVE
A
A
AIR SWIRL
STRATIFIED FUEL AIR MIXTURE
AIR ONLY
FLAME PROPAGATION
INTAKE STROKE
COMPRESSION STROKE (INJECTION)
COMPRESSION STROKE (IGNITION)
EXPANSION STROKE (TDC)
AIR SWIRL
EXHAUST
INTAKE
SPARK PLUG
INJECTOR
SECTION A-A
AIR SWIRL
INERTIA
SPARK PLUG
FUEL DROP
RESULTANT
DRAG
FUEL INJECTOR
AIR FUEL RATIO
≤15:1
21:1
28:1
35:1
42:1
∞:1
STRATIFIED FUEL-AIR MIXTURE
CENTRAL LOCATION OF THE FLAME PATTERN
AIR ONLY - NO COMBUSTION
SECTION A-A

the centre of the combustion chamber so that the fuel is concentrated at the ignition point. Upon ignition the flame spreads outwards radiallv towards the cylinder walls, though under part-load conditions it would not travel all the way because in the outer areas there would be only air present – and this is why the stratified charge engine is not sensitive to fuel octane number: there is little danger of spontaneous ignition of mixture remote from the source of the ignition.

There are some noteworthy advantages of the stratified-charge engine. The part-load fuel economy and reduced toxic emissions are the most obvious, but since knocking is not a problem it follows that the compression ratio can be increased indefinitely without regard to the octane number of the fuel. Since anti-knock lead additives can be dispensed with, spark plug fouling and chamber deposits will also be reduced, and the way is opened up to use a simple catalytic converter to eliminate any unburned hydrocarbon remaining in the exhaust (for example from lubricant finding its way into the combustion space). Perhaps most intriguing af all, the engine can make effective use of supercharging: all the usual objections to this are overcome, and thus we are brought back to the very beginning of this chapter and to a revision of its opening statement: if God had meant engines to be supercharged, perhaps He would have shown us the way to computerise variable injection and ignition timing.

The fact that Ford as well as Toyota, Isuzu (who are in collaboration with General Motors), and Chrysler, have bought development and production rights to the CVCC principle of Honda, suggests that the hopes of divine intervention could not long be sustained. The Honda engine achieves charge stratification without fuel injection, relying instead on a compound carburettor and an auxiliary inlet passage for the rich mixture to a tiny supplementary inlet valve which admits it to a precombustion chamber very much like that of prechamber compression-ignition engines.*

The Honda arrangement is evidently quite good enough to secure adequate stratification of the charge, the bulk of the air being ingested through a conventional inlet port so that each cylinder is served by a total of three poppet valves controlled conveniently by an overhead camshaft. Ignition in the rich mixture chamber is ensured by an ordinary spark plug of admittedly fairly wide heat range, and its timing is elaborately controlled. The outwardly conventional mechanical ignition distributor contains not only the usual centrifugal governor but also modulators that sense intake manifold pressure and coolant temperature.

* The same basic idea had been proposed a little earlier by the Azure Blue Corporation in California: their precombustion chamber was built into a special spark plug.

Courtesy: BP Trading Ltd.

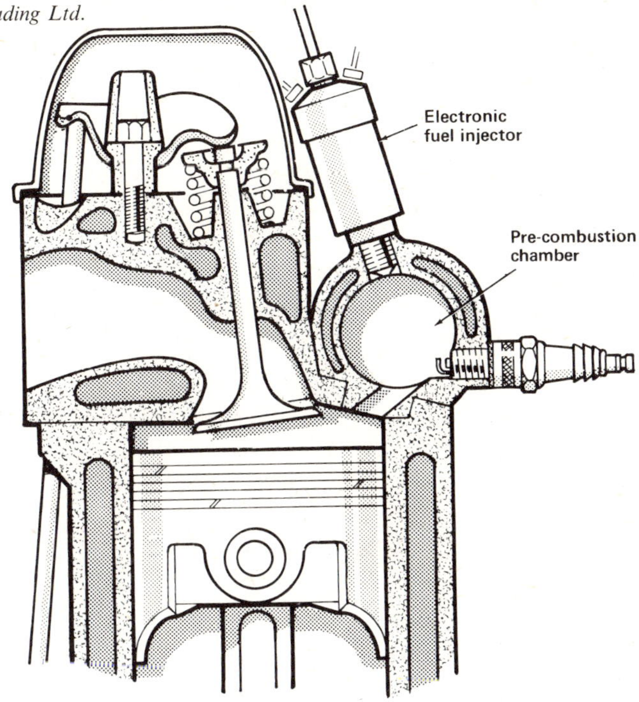

Although both Honda and General Motors have collaborated on stratified charge development, the direction taken by the latter differs from Honda's approach. In the General Motors' design, fuel injection is employed in the introduction of the mixture to the pre-combustion chamber.

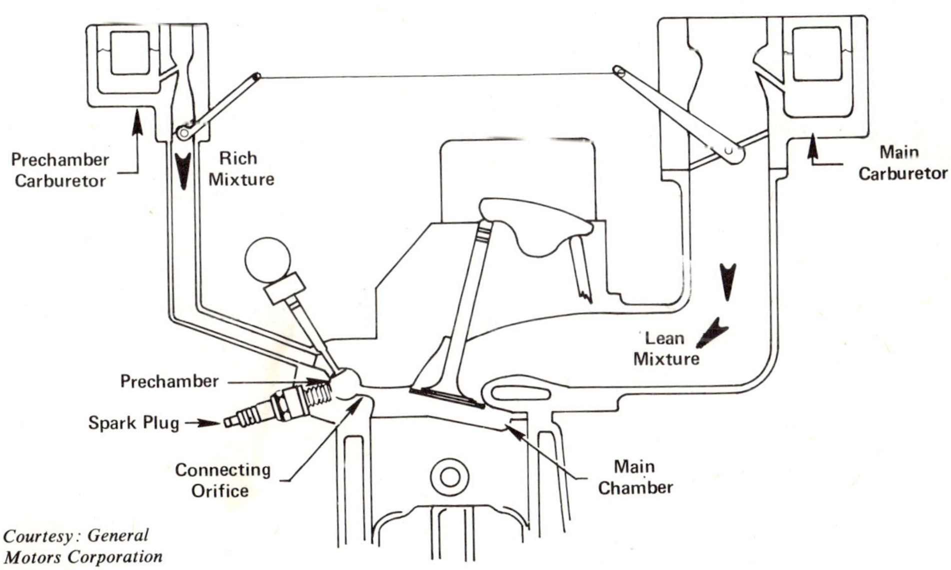

Courtesy: General Motors Corporation

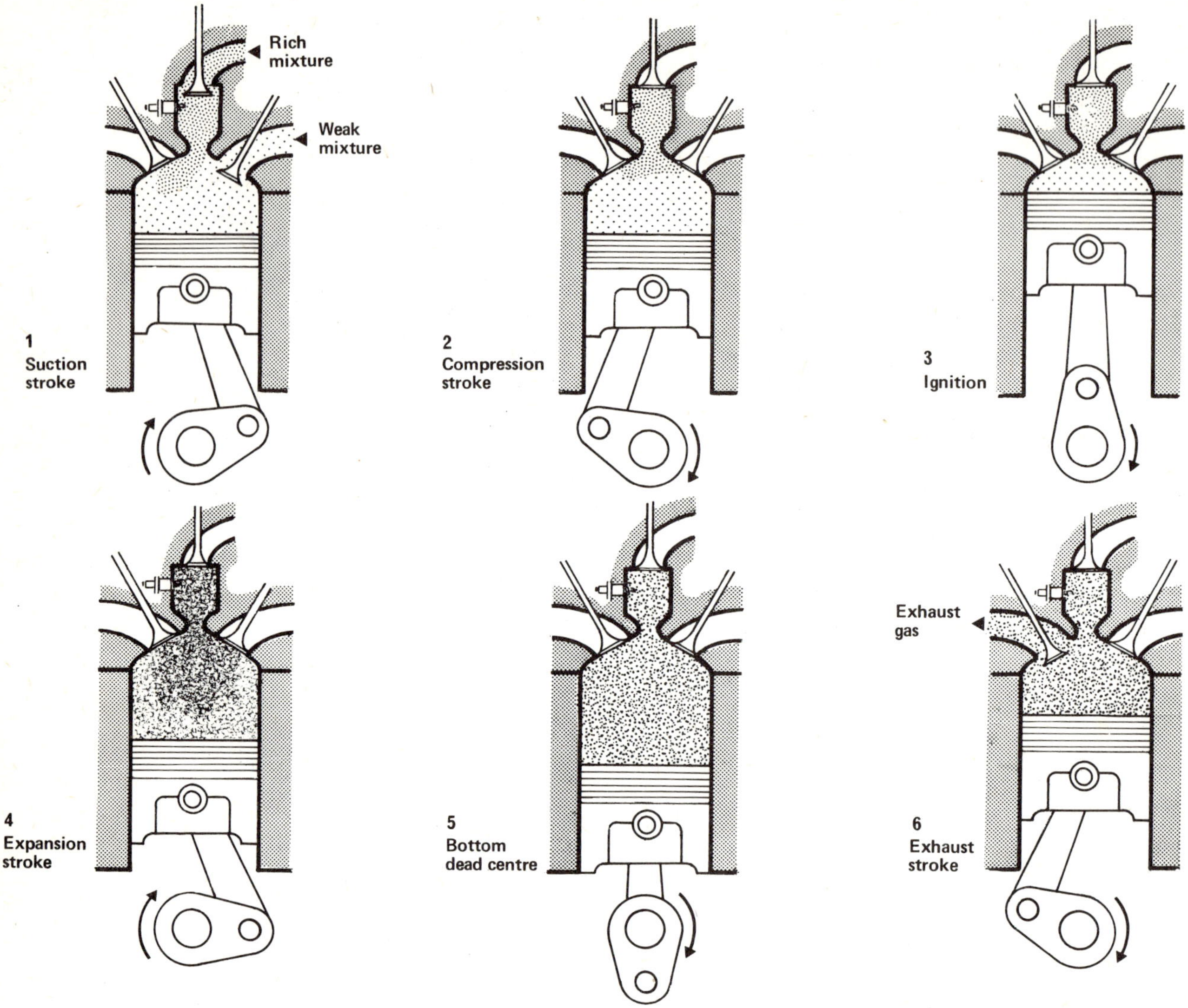

Eschewing fuel injection, Honda rely upon a compound carburettor and an auxiliary inlet passage for the rich mixture to a tiny supplementary inlet valve (seen clearly here) which admits it to the precombustion chamber in their CVCC stratified charge design.

Courtesy: BP Trading Ltd.

Apart from relying on carburation rather than injection, the feature which distinguishes the Honda CVCC engine from the classical versions of the stratified charge system exemplified by Ford is that there is little or no reliance on swirl. The initials were originally conjured to represent Compound Vortex Controlled Combustion; but in an early stage of development Honda learned that strong turbulence and vortex affects were neither necessary nor desirable in securing effective stratification. The straitened passage between the precombustion chamber and the main wedge-shaped combustion chamber in the cylinder head is evidently enough to prevent the mingling of the separately induced rich and weak mixtures into an homogenous one. Here indeed is an engine that is certainly not to be characterised as a might-have-been, nor even as a might-well-be: quite simply it is – and in view of the production resources of the companies now taking an interest in it, we may not for long be able to consider it unusual.

2. Engines of Unusual Construction

The introduction by Walter Hassan to his paper on the Coventry Climax 16-cylinder racing engine ended with a telling sentence: 'I will also discuss some of the shortcomings, as in fact one learns more from one's mistakes than from the successes.' This might have been the underlying theme of the whole meeting devoted to unusual engines, just as it is the justification advanced by many engineers for getting involved in motor racing – an activity in which only those who make their mistakes early and learn from them quickly can hope to remain competitive.

The background to this Climax flat 16 was quite straightforward. The firm produced a very conventional $1\frac{1}{2}$-litre V8 to suit the Grand Prix racing formula that took effect in 1961. In the following year they shortened its stroke by 16 per cent, enabling the crankshaft rotation to be increased by nearly 15 per cent and the power by 9 per cent; and with some other refinements this kept the engine tolerably competitive until the end of 1963. By this time it was clear that the competition was catching up, so Coventry Climax embarked on two parallel projects. One was considered a stop-gap, involving drastic shortening of the stroke so that it was only 76 per cent of the original, together with a four-valve arrangement in pentroof combustion chambers replacing the twin-valve hemispherical ones of the earlier design. The other project was to design a completely new sixteen-cylinder engine. The reasons for both were similar in that, as bmep appeared to defy substantial improvement, an increase of revolution rate seemed to offer the only chance.

As it happened, the four-valve project got off to a poor start, a lot of precious time passing before it was realised that the ignition would have to be considerably advanced to make the engine work satisfactorily. In fact its power was just a little higher, but its torque was appreciably better and the much higher rate of revolutions allowed a similar reduction in overall gear ratios which gave better acceleration still; and this was enough to make the engine very competitive indeed.

Pursuit of this same reasoning led Hassan to choose the sixteen-cylinder engine because an extremely short stroke of 1·6 inches was possible without the need to have very much over-square cylinder proportions – which at the time of the design seemed to give headaches with the ultra-short-stroke V8. He was also put off the four-valve configuration for similar reasons. The flat sixteen, with a bore of 2·13 inches, promised a safe running speed in excess of 12 000 rev/min without exceeding conventional limits for piston velocity.

The long crankshaft of a straight 8 or a flat 16 is particularly subject to torsional vibration problems, though they are perhaps less peculiar than those affecting an in-line 6. To minimise these the engine was designed around two single-plane crankshafts set at 90° to each other, with a pinion gear with tapered stubs coupling them together. The taper was very small, about 2 per cent, and assembly relied on the SKF oil injection process. This was one of the most elegant devices in the whole design, being admirably simple and cheap, yet effectively solving what is normally a difficult design problem. The arrangement was that the tapered stubs projecting from each flank of the central drive gear fitted into corresponding recesses in the end of each half-crankshaft, and the female tapers were expanded by injecting oil at high pressure through an access hole in the crankshaft itself. With the female taper thus expanded the components could be pulled up together, trapped air finding its way out of a release hole conveniently drilled in each crankshaft half. Dismantling the assembly was even easier: the usual hand pump was coupled to the access holes and pressure built up, whereupon the tapers would part and the components would gradually separate themselves, in effect being blown apart by the high-pressure oil.

Perhaps the only other constructional features worthy of notice were the gear drives and the disposition of necessary auxiliaries. Power was taken from the central pinion to a gear and drive shaft situated below the crankshafts, reducing the output speed by 25 per cent and thus permitting the retention of the existing transmission in the cars for which the engine was intended. Another gear lower down was used to drive the lubricating pressure and scavenge pumps, with the driven gear and spindle in the pressure pump driving the coolant pump at a lower level while camshaft and auxiliary drives were taken from the same central crankshaft pinion upwards. The auxiliaries were a problem; there were no 16-cylinder versions available, so two 8-cylinder fuel injection pumps had to be used and two 8-cylinder ignition distributors. These together with an alternator were housed either side of the central gear case over the top of the crankcase and between the steeply inclined inlet trumpets, thus producing a very compact arrangement. Even with these double quantities of fuel and ignition components the engine

The flat-16 Coventry-Climax GP engine was originally conceived by Walter Hassan as a replacement for the $1\frac{1}{2}$ litre GP Coventry-Climax V8 at the end of 1963. Its configuration was largely the result of Hassan's development work on the V8 where it was found that by shortening the stroke and increasing the crankshaft rotation, a considerable increase in power could be achieved. The flat-16 permitted an extremely short stroke of 2·13 inches without the need for very much over-square cylinder proportions but in the event proved disappointing. Although problems with the V8 were subsequently overcome, it was at about this time that Coventry-Climax decided to withdraw from racing.

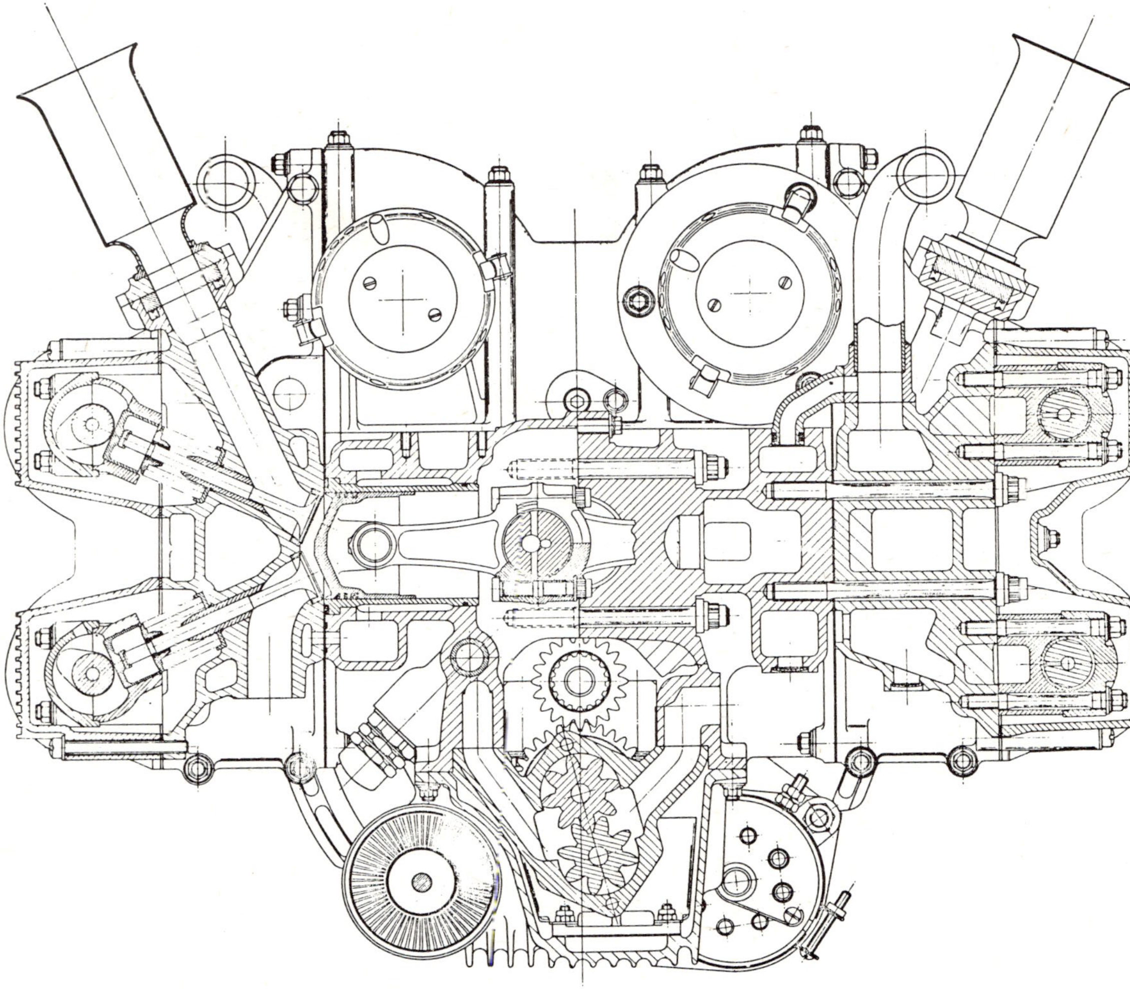

A vertical section through the 1963/4 Coventry-Climax flat-16 GP engine.

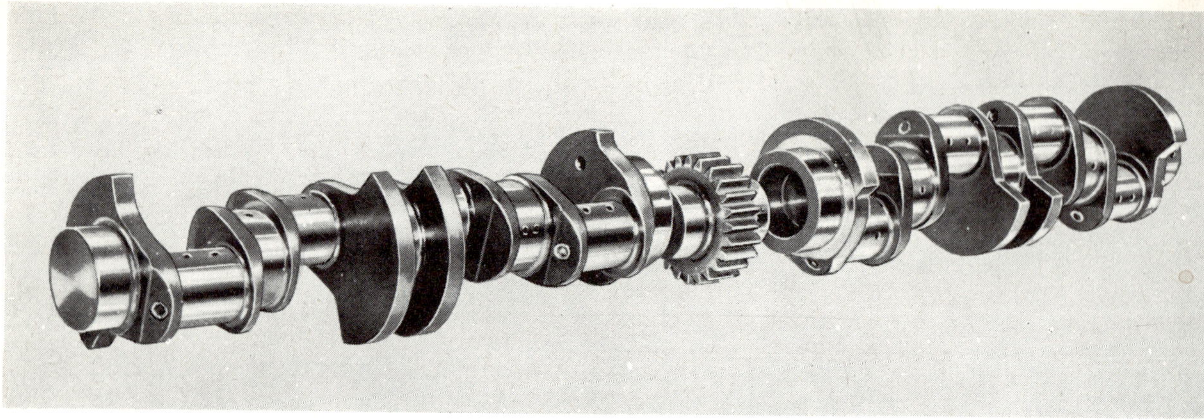

The crankshaft of the flat-16 Coventry-Climax racing engine employed an eight-throw crankshaft (four throws on each side of a centrally positioned gear which meshed with a larger gear on the output shaft). Each set of throws lay in a single plane but the two halves of the crankshaft were phased 90 degrees apart. Torque transmission was dependent upon the degree of interference between the mating surfaces of the gear boss and the recesses in the relevant ends of the half shafts. Attempts to use mechanical force and thermal shrinkage to obtain the necessary interference during assembly were unsuccessful, but the problem was overcome by the employment of the SKF oil injection method, and axial displacement and fine adjustment were thus possible. Details of the assembly are shown below, and the air release hole and pressure oil duct (one of each was machined in each half of the crankshaft) will be noted.

Courtesy: SKF (UK) Ltd.

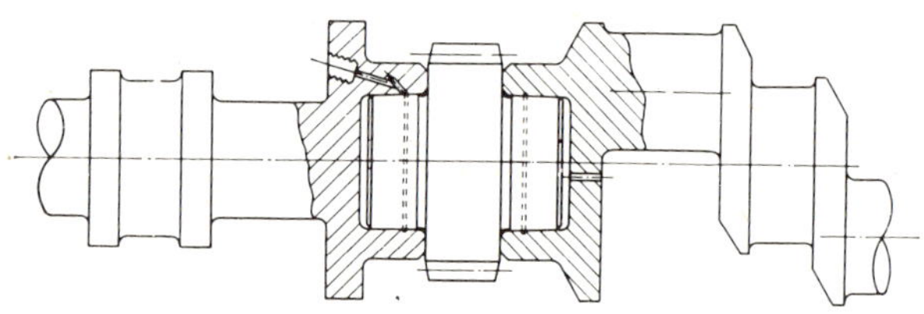

weighed only 10–15 lb more than the previous 8 cylinder engine and fitted into the same wheelbase, largely because of the extremely compact auxiliary drives.

As a typical piece of motor racing jewellery the flat 16 engine which I saw on the dynamometer bed, and also in pieces, very early in 1965, was as pretty as any. The results it yielded were disappointing, though. Work ceased on it when the four-valve engine was persuaded to give satisfactory results, and at this stage the performance had only reached 210 bhp at 12 000 rev/min. There were difficulties in three respects. The initial run showed the simultaneous firing imposed by the horizontally-opposed cylinder arrangement in conjunction with side-by-side connecting rods on four-throw or 180° cranks caused a torsional vibration in the final drive shaft at low crankshaft speeds. These were below normal operating rates, so the problem was not terribly critical and was overcome by increasing the torsional stiffness of the drive shaft and evading any running below the critical speed which was around 3000 rev/min. This problem having been overcome, another became apparent: oil build-up in the gear casing caused a loss of power and shortage of oil at the pick-up for the pump. A lot of fettling and much fiddling with deflector plates eventually reduced the losses to a point where suitable power was achieved, but then some valve spring defects were beginning to show. This was attacked in two ways: first by designing springs with better characteristics, and secondly by designing cam profiles which offered slightly less severe valve acceleration. It was only after this that the engine could be run up to 12 000 rev/min, and achieve 210 bhp – which was scarcely enough.

Morse tests showed the indicated horsepower to be as expected and that frictional or pumping losses were greater than estimated. At 9000 rev/min the 16-cylinder engine showed a mechanical efficiency of about 79 per cent, similar to that of the V8. Tests on the power required to operate the valve gear of the four-valve V8 showed this to be 6·2 bhp at 10 000 rev/min, rising to 7·7 bhp at 11 000 rev/min which was a pretty steep increase. The rig available was unable to run the 16-cylinder valve gear up to 12 000 rev/min but it was evident that the power required was more than double, say 14 bhp. It was also known that the oil drainage problem was not completely solved, so the target figure of 230 bhp was perhaps not so very far out of reach at the time Coventry Climax abandoned the engine.

The conclusion drawn by Hassan from all this was that some worthwhile power at high crank rates could come from quite small improvements in mechanical efficiency – bearings with less oil drag, perhaps of the anti-friction type, with better attention to oil return paths and reduced hold-up back to the scavenge pumps, perhaps extra scavenge facilities from troublesome areas . . . and certainly with hindsight, which he described as ‘a most useful commodity', he thought that four valves per cylinder would have made a vast difference to the results. As it was, the engine was dropped and it was not long afterwards that Coventry Climax withdrew from official participation in motor racing. In 1966 a new formula came into effect permitting engines of 3 litres displacement, and the company was reluctant to start all over again.

Nevertheless I found myself discussing this engine some years later with Mr Windsor-Smith, then engaged in the Guy/Daimler branch of the Jaguar group in which Coventry Climax played an important part. He pointed out that the rules permitted supercharged engines of $1\frac{1}{2}$ litres to compete with the unblown 3 litres, and told me that he had been very keen to develop the 16-cylinder Climax accordingly. According to his calculations, it would have been feasible to supercharge the engine so as to develop 400 bhp on the mandatory pump petrol. This would have been a surprising achievement, in view of the limitations of boost pressure imposed by the fuel; but had it been encompassed, the engine would almost certainly have been extremely competitive in the early years of the formula. Compared with the opposition it would have encountered, it would have been gratifyingly light and compact, not to mention being more powerful than practically all the rivals available to it in 1966. Whether the lightness would have been offset by the increased fuel load necessary to serve the thirst of a supercharged engine is open to some doubt, but at least this supplementary burden would be one that could be shed progressively during a race. Keeping it cool might have been another difficult problem.

Closely engaged with Walter Hassan in the design and development of the flat 16 was Harry Mundy, who later went on to apply the lessons learned to the design of an engine that might prove highly competitive under the 3-litre rules. Although enjoying some collaboration from Hassan, Mundy produced this design more or less as a piece of private enterprise, either in the hope that Jaguar might adopt it (a hope which was denied fulfilment) or as a piece of inspired flag-waving that was nicely timed for the Institution discussion. His review of the background made an interesting start: of the engines in current use there was one V8, one flat 12, and a pair of 60° V12s; and there appeared to be little difference in performance between them, the maximum power being in the region of 435–455 bhp. So much is known today, said Mundy, about inlet port shapes together with individual air intakes furnished with fuel injection, that more work on them could be expected to give little hope of improved power from this source. In fact there is only likely to be a marginal improvement in power resulting from frictional savings, an increase in operating speed and, he thought, exhaust

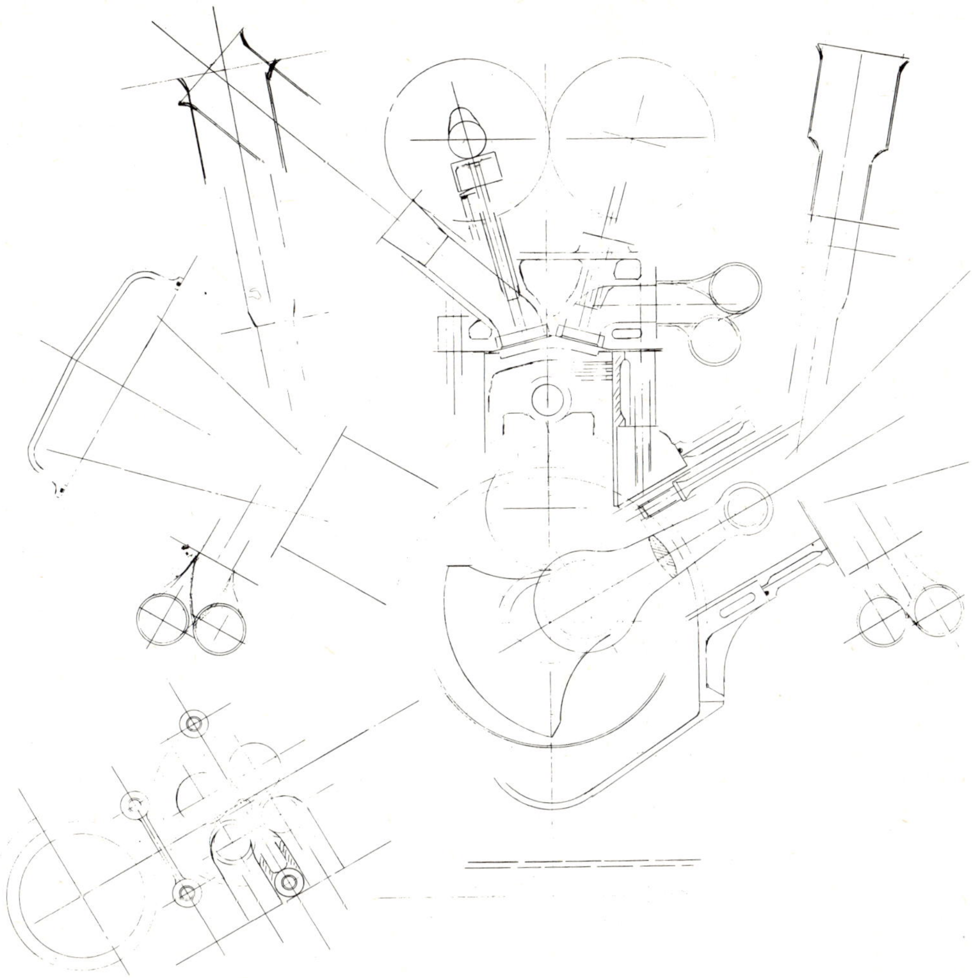

The Mundy Trident or 'broad-arrow' design advocated by Harry Mundy was largely the result of his experience during exhaust tuning development with Coventry-Climax. Concentration on exhaust tuning had proved that spectacular improvements in maximum power and torque could be achieved with clusters of pipes from four cylinders: the Trident was simply three in-line fours.

Courtesy: H Mundy.

pipe tuning. This approach, together with the reduction of engine weight, seemed to be the most fertile region for improved vehicle performance.

The part played by exhaust system design was strongly emphasised. Coventry Climax, as Mundy pointed out, were able to make quite spectacular improvements from exhaust tuning – not only to maximum power but (equally important) to the shape of the torque curve. These improvements were made on four and eight cylinder engines with exhaust pipes made confluent in groups of four. From observation as well as from experience with the 5-litre twin-cam Jaguar V12 prototype engine, the same response would not be obtained with tuned groups of three pipes.

This response to exhaust tuning was the main reason for putting forward the proposals for a 'trident' or broad-arrow layout, with its three banks of four cylinders placed at 60°. It provides equal firing impulses and the normal firing sequence of a four-cylinder engine in each bank: thus the exhaust can be linked as 1–4 and 2–3 and converge into a single outlet on each bank, or it can have four pipes of equal length converging to a single outlet. Moreover the balance characteristics of a broad-arrow twelve are good: such an engine does suffer from secondary shaking forces, but of the two alternative arrangements possible by adjustments of the firing order, one would give the engine greater freedom from vibration than enjoyed by the 90° V8 engine with its single-plane crankshaft which by that time was used in the great majority of Grand Prix cars. Mundy was referring of course to the Cosworth Ford V8, though he was reluctant to spell the name out; but in any case the ability to produce an engine vibrationally superior to the Cosworth is no great thing to claim. A 90° V8 with a flat crankshaft resolves the secondary shakes from each bank into an horizontal one manifesting itself at twice the rate of crankshaft rotation – and of course it was Coventry Climax who set the fashion, adopting crankshafts of this nature in the $1\frac{1}{2}$-litre V8s in the early 1960s. This was to simplify the clusters of exhaust pipes that, with the two-plane crankshaft, required to be crossed over in a space-consuming style that made them look more like an ecstatic squid than a piece of engineering.

There were other advantages in the Mundy Trident layout. The shorter overall length (compared with other 12-cylinder configurations) give greater freedom in chassis layout, particularly in respect of space for fuel tanks. The engine was expected to be not more than 20 inches long and to weigh about 325 lb;.and this might constructively be compared with the figures for the Cosworth V8 (21·5 inches and 358 lb) and the BRM V12 (29 inches and 375 lb). The Trident figure was nevertheless an approximation, for the engine had not been designed down to its last detail. Much of it was in accord with current fashion, including the four-valve head layout with the valves at an included angle of 30°, the stroke/bore ratio of 0·646 and the mean piston velocity of 3949 ft/min at the maximum envisaged crankshaft rate of 11 800 rev/min. Mundy estimated that an ultimate power of 460 bhp might be developed, which would correspond to a peak power bmep figure of little more than 170 lb/in^2.

Perhaps that is why the engine was never built; the figure hardly bears comparison with the best achieved by the opposition, let alone the best achieved by Coventry Climax in the past.

Constructionally the engine is interesting in its bottom half. Emphasising his view that the greatest scope for improving the power output of an engine lies in the elimination of mechanical losses as well as in exhaust tuning, Mundy stipulated one-piece elliptical-section connecting rods side by side on the crank pins of a built-up crankshaft. In a broad-arrow twelve with a single-plane four-throw such as he proposed, and exploiting the SKF oil injection method for assembly as in the Coventry Climax 16 cylinder engine, such a construction would require only five crankshaft components. A valuable attribute of such a construction is that it enables a one-piece connecting rod to be used with either plain or rolling-element bearings: using caged rollers, a crankpin of 1·5 inch diameter was calculated to be adequate for the anticipated loads, but in the interests of increased crank stiffness lead-bronze sleeve bearings would be possible and would enable the pin diameter to be increased to 1·625 inch without increasing the big-end width. The rolling-element bearing is inherently unsatisfactory for a big end because of the angular accelerations to which the rollers are subjected, and one might prefer plain big ends for this reason; but no such objections apply to the main bearings, where rollers would reduce the demands on oil pumping and thus minimise the losses that the designer is so keen to avoid. Mundy was not sure what to do, suggesting that lead-bronze bearings would probably be used for the centre three mains and a combination of ball and roller bearings for the front and rear.

The only other interesting feature of the design was the need for the connecting rod length : stroke ratio to be as great as 2·25 : 1, in order to assemble the pistons. Although he did not mention it, the designer was doubtless content with this ratio because of its beneficial effect in reducing maximum piston accelerations. Even with this aid the maximum piston acceleration in the Trident would be 156 100 ft/sec^2 at peak revs, a figure 19 per cent higher than in the 16 cylinder Climax engine, and only negligibly lower than in the 1½-litre 16-cylinder supercharged BRM that, a few years earlier, had constituted a warning to designers of the perils of letting their enthusiasm for high revolutions run away with them.

The Mundy Trident makes an interesting comparison with the morphologically similar Napier Lion, drawings of which were adduced in evidence. The Lion had rolling-element main bearings and plain big ends, with one master connecting rod and two articulated slaves to each crankpin. It was not particularly compact longitudinally, but in its intended aviation application it had no particular need to be. On the other hand its expansive frontal area was an aerodynamic handicap, and Mundy pointed out that it was this that probably prevented the format from being followed by other aviation engine designers despite the engine's lower weight than a V12 of comparable displacement or power. The Lion had one constructional feature missing from the Trident but present in the Mitsubishi tank engine discussed in chapter 1: the upper cylinder head deck was attached by the valve seats being threaded into it. In the case of the Lion, which had one-piece forged cylinders closed at the top, it was a matter of fixing the entire head to the top of the cylinders; in the V10 Mitsubishi the cylinder head consists of two major component parts, a steel combustion chamber as the bottom half and an aluminium alloy part above it, joined together by the screw threads of the exhaust valve seats. The head itself is fitted to a steel cylinder by a shrink-fit screw thread. The steel nether portions of the head were cooled by air through the surrounding fins and also positively cooled by the lubricating oil which circulated through machined oil passages located between the valve ports and the fuel injector inlet.

The effective cooling of finned air-cooled cylinders has often led engineers up some unfortunate alleys. A notorious example was the ABC Dragonfly engine, designed to meet a British Air Board specification issued in 1917 at the instigation of the Admiralty, who wanted a new fighter engine that would develop at least 300 hp. They specified an air-cooled radial not exceeding 42 inches in diameter, which encouraged three firms to propose two-row 14-cylinder radials while ABC put up a single-row design embracing 9 cylinders.

This was the Dragonfly, and its design – not to mention the salesmanship of its designer Granville Bradshaw – quite captivated the Government. Bradshaw had already submitted a 7-cylinder radial called the Wasp, an engine that behaved very well in its initial tests; and with this admittedly good engine bearing witness to his competence (he had already proved himself a gifted motorcycle designer in the immediate pre-war years) it was assumed that his 9-cylinder Dragonfly would be equally meritorious. On paper – which was as far as it then went – it developed 340 hp and weighed little more than 600 lb. What made the design attractive was the number of clever ideas that would simplify production; but what made it dubious was

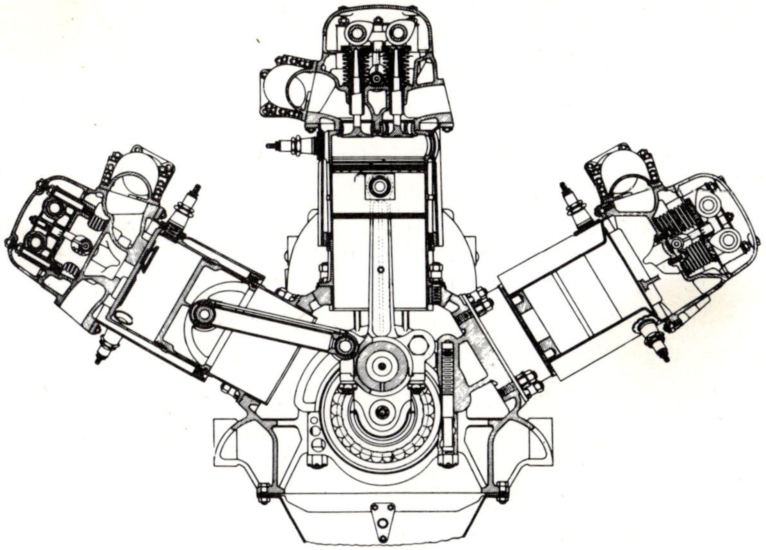

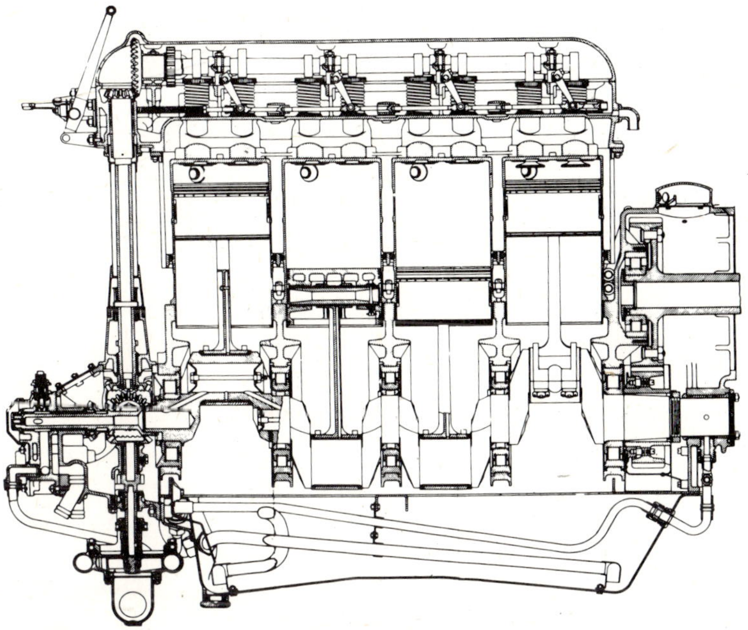

The Napier Lion 24-litre aero engine introduced in 1918 was unique in its field and one of the most successful. Like many of Napier's offerings it combined apparent complexity with excellence of design and manufacture, and as raced in the Schneider Trophy aircraft it developed 875 bhp for a weight of only 930 lb—a very respectable power-weight ratio—and its specific fuel consumption of 0·32 lb/bhp/h has probably never been bettered. It is interesting to contrast it with the Mundy Trident design.

Courtesy: Science Museum, London.

Only the provision of a better view for the pilot, or a reduction in frontal area, might justify the disposition of the 1912 Canton Unné 110/130 hp radial aero engine with its vertical crankshaft and angled drive, a necessary corollary of which was water cooling for the cylinders—to which another French maker of radials (Salmson) later became addicted. Salmson also employed another feature of this Canton Unné that was much more endearing in principle if ruinous in practice: instead of master and slaves among the connecting rods, their disparate motions provoking secondary vibrations, there was a planetary gearset that served to prevent disturbance by making them *all* slaves. Such egalitarianism could not prosper even in its maternal republic.

a number of other curious ones that Bradshaw insisted, in open defiance of the scientists of the Royal Aircraft factory at Farnborough, would promote excellent performance. One of these was his idea of copper-plated steel cylinder finning for supposedly effective cooling. The Government and the aircraft industry went wild with enthusiasm, and the Ministry of Munitions ordered that production of practically every other engine (save two Rolls-Royce and the BHP/Siddeley Puma) be stopped by 1919 so that all manufacturers could standardise on quantity production of the Dragonfly. Perhaps the experience taught them (and perhaps it could teach us) a lesson, for the Dragonfly turned out to be disastrous. It weighed half a hundredweight more than ABC predicted and developed 45 hp less; and it had an irredeemable tendency to break itself into pieces after only a few hours in the air. About twelve hundred were built by the time it went into full production in 1918 before the Establishment realised the error of its ways and abandoned the thing.

Quite the most extraordinary system of cooling was that proposed by Ernest Chatterton for his Simplic 2-stroke engine. It had no fins at all but consumed about five times more air than it needed for combustion, employing the bulk of the air to cool the engine from the inside just where it had been made hot, rather than having to divine the heat at a distance down the temperature gradient of cylinder walls and fins. What prompted Chatterton to work out this idea was his distaste for the complicated design of supercharged engines, particularly those using intercoolers for the intake air, and the general unsuitability of such a principle for engines intended to generate less than about 150 bhp. There is no doubt that high efficiency is very difficult to obtain in small turbochargers, due to high gas losses, and the cost of a turbocharger is not much decreased with its size. Arguing properly (and if only others would agree) that power:weight ratio is a more important parameter of engine performance than power:displacement ratio, Chatterton concentrated on the attainment of low weight and small engine dimensions in a construction of the utmost simplicity.

In principle the Simplic was a two-stroke engine with uniflow scavenging through inlet ports in the cylinder wall and a poppet exhaust valve in the head. However, the engine requires no blower for charging nor any fan for inducing a flow of cooling air. All this is made possible by the engine operating with a very high expansion ratio, no less than 60:1. There have been a number of engines exploiting the more-complete-expansion cycle, in which the expansion ratio is greater than the compression ratio: the idea makes good thermodynamic sense, but seldom has it been exploited with such ingenuity as in the Simplic. So great is the expansion in Chatterton's cylinder that, when the inlet or scavenging ports open in the cylinder wall, the gas expansion has already gone so far that a depression of 0·035 to 0·05 atmospheres below ambient is present in the cylinder. As a result, air from the atmosphere begins to flow into the cylinder through the bell-shaped inlet chest surrounding it; and the air stream is directed by the ports towards the piston crown, where violent swirl is promoted. Just before the piston completes its downwards stroke to bottom dead centre, the exhaust valve opens and exhaust gases are led into the exhaust pipe. Since however there is still a slight vacuum in the cylinder, a higher vacuum must be produced in the exhaust pipe – and this is accomplished by connecting two cylinders (with explosions displaced by 180° of crank angle) into a single branch of the exhaust pipe.

During the upward travel of the piston, airflow from the atmosphere to the cylinder continues, and the cylinder is scoured by fresh air until the closure of the piston-controlled ports at about 55° past bottom dead centre. In the final exhaust stage, until the closure of the exhaust valve 115° past bottom dead centre, exhaust gases in the pipe must have a velocity sufficient to develop a depression to scavenge the second cylinder whose exhaust valve will open 50° after the closure of the valves of the first cylinder.

Thus it may be seen that only a small proportion of the cylinder volume is used for combustion, and this is determined (as is the compression ratio, which is no different from what normal engines feature) by the timing of the exhaust valve closure. Until this occurs the engine is being cooled internally by the large quantity of air ingested and passing through it. Due to the high expansion ratio, the mean temperature of the cycle is low and the cylinder wall is exposed to high gas temperatures only for one-third of the time appropriate to a normal cycle.

Delayed closure of the exhaust poppet valve is the accepted way of contriving a more-complete-expansion cycle in a two-stroke engine. In a four-stroke the early closing of the inlet valve achieves the same effect by giving an effective compression ratio lower than the expansion ratio. However, most engines working on this cycle have been conceived as compression-ignition types, for in practice maximum cylinder pressure is limited by considerations of stress, whereas in spark-ignition engines it is limited by detonation. The advantage of the cycle is the possibility of an efficiency higher than could be obtained with an expansion ratio equal to the compression ratio; but the disadvantage is considerable, amounting to a mean effective pressure lower than the conventional arrangement with the same maximum pressure. In the Simplic it is only about 40 or 50 lb/in^2, though the lightness and compactness of the engine makes it feasible that an aluminium V4 weighing 220 lb might produce 40 bhp fairly reliably. It must be stressed, however,

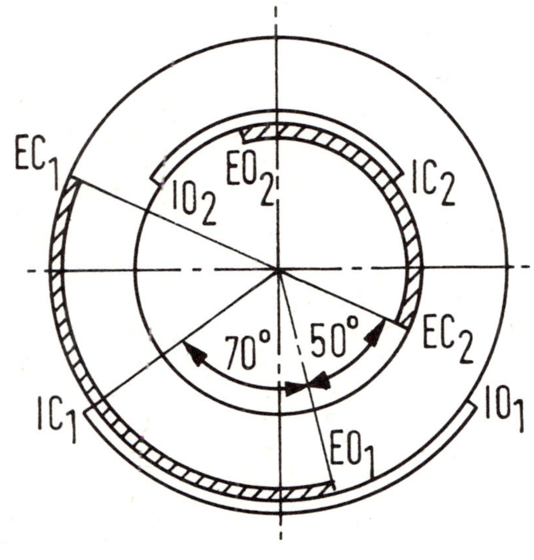

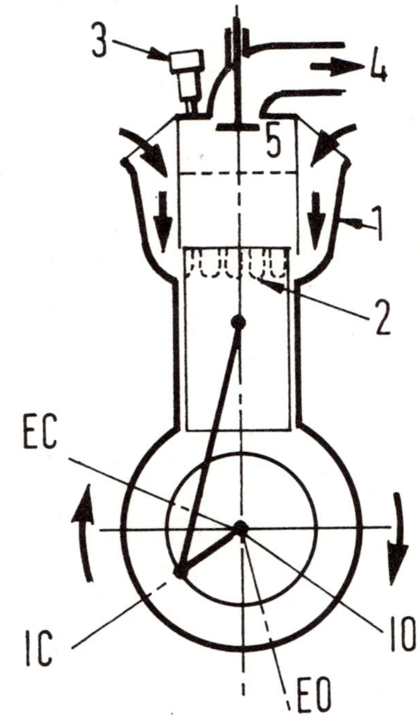

With an expansion ratio no less than 60:1 and a compression ratio sensibly lower, the Chatterton Simplic was one of numerous hyper-expansion engines. Evacuation through mechanically-operated exhaust valves into coupled pipes from two cylinders phased 180 degrees apart induced a flow of air into the cylinders—a flow five times greater than was needed for combustion, the excess serving to cool the engine internally. It is an intriguing (and apparently successful, on a pilot scale) way of harnessing the two-stroke cycle to give modest power in an exceptionally light engine.

that the more-complete-expansion cycle is practical only for engines that are not frequently operated at light loads. The mean cylinder pressure during the latter part of the expansion stroke tends in light-load operation to be near to or even lower than the friction mean pressure, and when this happens the hyperexpansion portion of the cycle may involve a net loss rather than a gain in efficiency. The principle might best be employed for engines such as are used in pipeline pumping stations where great reliability, long life, and good fuel economy, are required. Running on a fuel such as methane, a large hyperexpansion engine might be a good proposition; but a large engine is not subjected to the restraints on turbocharger efficiency that prompted the design of the little Simplic. It is not difficult to see applications for the latter driving small pumps, generators, and the like.

Nor is it difficult to resist the argument that these considerations should have been discussed in the chapter on unusual charging and scavenging. What, after all, is the turbocharger or the exhaust power turbine if it is not a more complete expansion device? All this must be accepted; but the unworldly simplicity and apparent crudity of construction of the Chatterton Simplic gives it a place here.

3. Engines with Unusual Valve Apparatus

The spring-loaded cam-operated poppet valve is a hideous contraption. It must inevitably be too big, too heavy, too slow moving and too sensitive to heat and fit to do half the jobs required of it, if it is to do the other half satisfactorily. For the greater part of the history of the internal combustion engine, it has constituted a barrier to the achievement of the higher power and greater volumetric and thermal efficiencies that have been demanded of engines; and even now it is no more efficient and no more reliable than most other components – which is not always saying a great deal. The fact that it has so long remained the established and most successful means of controlling the intake and exit of fresh charge and vitiated gases probably tells us more about the habits of thought of the engine industry than it does about the valve itself. It was inescapable that every imaginable alternative device should have been tried; but whilst many conventional valve arrangements have been pretty disastrous, the alternatives have only contributed eventually to the history of failure. Few aspects of engine design better illustrate the observation by Samuel Smiles that *We often discover what will do by finding out what will not do; and probably he who never made a mistake never made a discovery.*

As it happens, that statement was made by Smiles in a book entitled *Self Help*; and it may be significant that this branch of engine design has been a happier hunting ground for the amateur than any other. Although the professional engine designer would be quick to read some pejorative insinuation in this, none is intended: there is little but good in the amateur, nothing to fear from the dilettante engineer. On the other hand there is little more dangerous than an enthusiast, and too many of the novelties of valve gear that have enlivened the otherwise dull affair of engine design patents have been the creations of uncritical enthusiasts. They are now no more than names in the records and occasional drawings in textbooks; but if they have done no more, they have drawn attention to a state of affairs that the regular year-in year-out professionals have been perhaps too ready to accept as inevitable.

Of all the engineers delivering addresses to the Institution at the discussion on unusual engines, only the two whose subjects were valve gear were not and did not represent men whose whole-time working preoccupation was with the internal combustion engine. Like all the others, both had tales of failure to tell; but unlike any of the others, one of them could point to a continued programme of development that showed promise for the future.

The important thing about an engine is not how clever it may be but how timely. Leonardo da Vinci's helicopter had to wait centuries before engines were available to make it feasible; and de Dion's rear axle had to wait thirty years for the increased power of American racing cars to make it desirable. The Aspin rotary valve is now more than forty years old – but as the internal combustion engine falters through the critical 70s, its future never more in peril, conditions have suddenly come to favour the adoption of Frank Aspin's brainchild. As luck would have it, he has been patiently developing the valve throughout the intervening decades so that it is now even better adapted for the future which holds so much promise for it.

History takes some ironic turns. The original Aspin valve was conceived – like so many others, including the Cross, Minerva and Tacconi rotary valves – with a view to eliminating the then troublesome poppet valve. Nowadays it is generally (if in some quarters reluctantly) accepted that the poppet valve works; but new standards and new shibboleths, new social pressures and ecological ideals, threaten to rob the internal combustion engine of the advances it has enjoyed. Engines that satisfy the new rules about noxious emissions are poor emasculated things; and if these rules are to continue unrepealed, the future presents a forbidding picture of low compressions, leadless fuel, poor idling, low thermal efficiency and a succession of dull and expensive engines all girt about with supplementary plumbing and each bigger and more gutless than its predecessor. If on the other hand the most recent panic over the threatened diminution and ultimate exhaustion of our fuel supplies takes precedence over the claims of the environmentalists, all we shall maybe see is a succession of weeny, weedy and weak contrivances humbly producing their pathetic powers at modest revolutions in return for a starvelling diet of grossly overpriced fuel. Or there is the Aspin: it will run at high compression ratios on the coarsest fuels, leave a clean exhaust, be more efficient, not less, and exhibit specific outputs such as a racing engine might, combined with flexibility at low speeds such as an old Silver Ghost should. For good measure, its fuel economy rivals that of a good diesel engine and production costs could also be cut because the mechanism allows automated engine assembly.

At the time of the Institution's meeting Frank Aspin was a 58-year-old

The first embodiment of the Aspin rotary valve idea consisted of a gear-driven vertical sleeve with square porting, rotating at half engine speed. Conceived in the early 1930s when its designer was still a teenager, its potential was limited by the size of the port.

Courtesy: F M Aspin

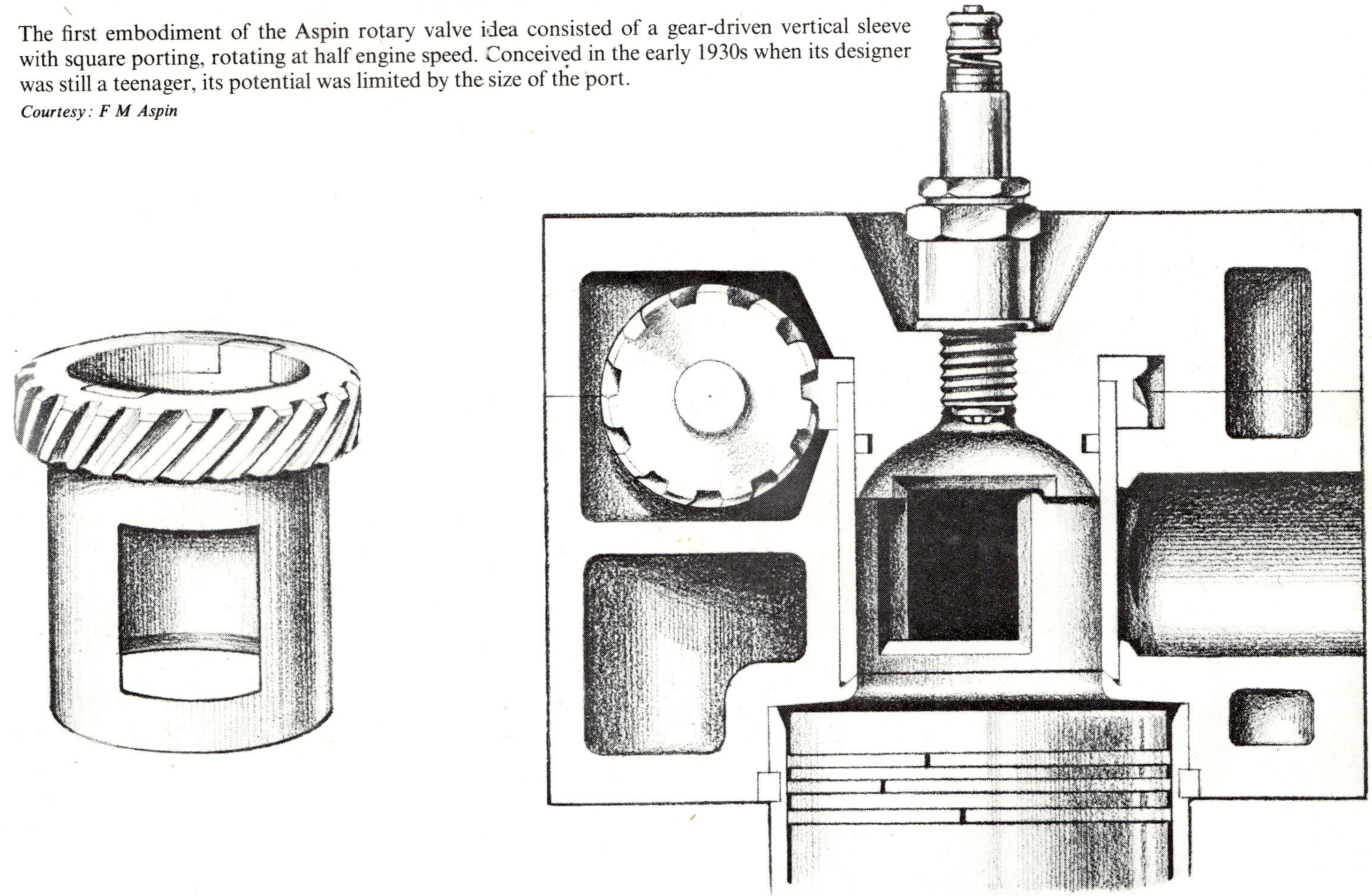

Courtesy: F M Aspin.

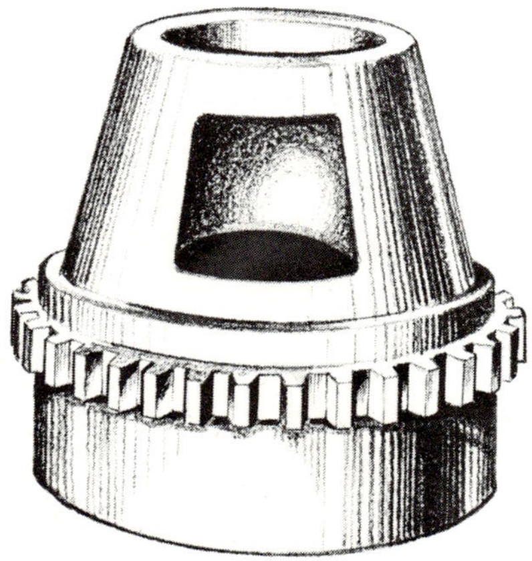

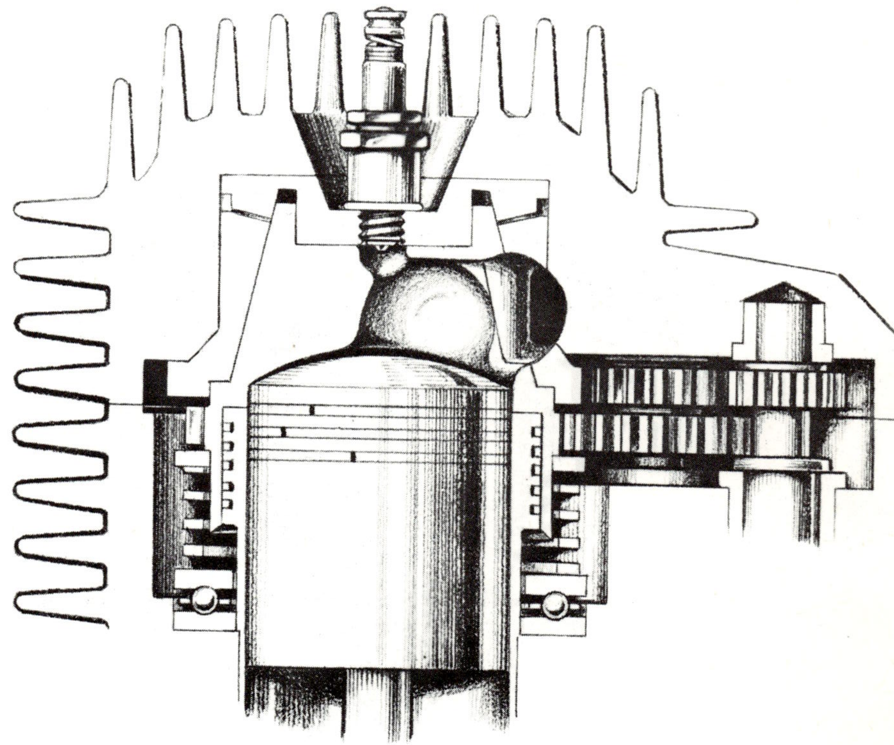

The third stage in the development of the Aspin valve, like the first, utilised a central sparking plug but differed in that it had now become a bottom driven rotor and the familiar conical shape of the later models had begun to appear. This design was successful from the point of view of performance, but was expensive, being made from case hardened nickel steel running on an aluminium bronze housing, and tended to seize if not thoroughly warmed up before being given full load.

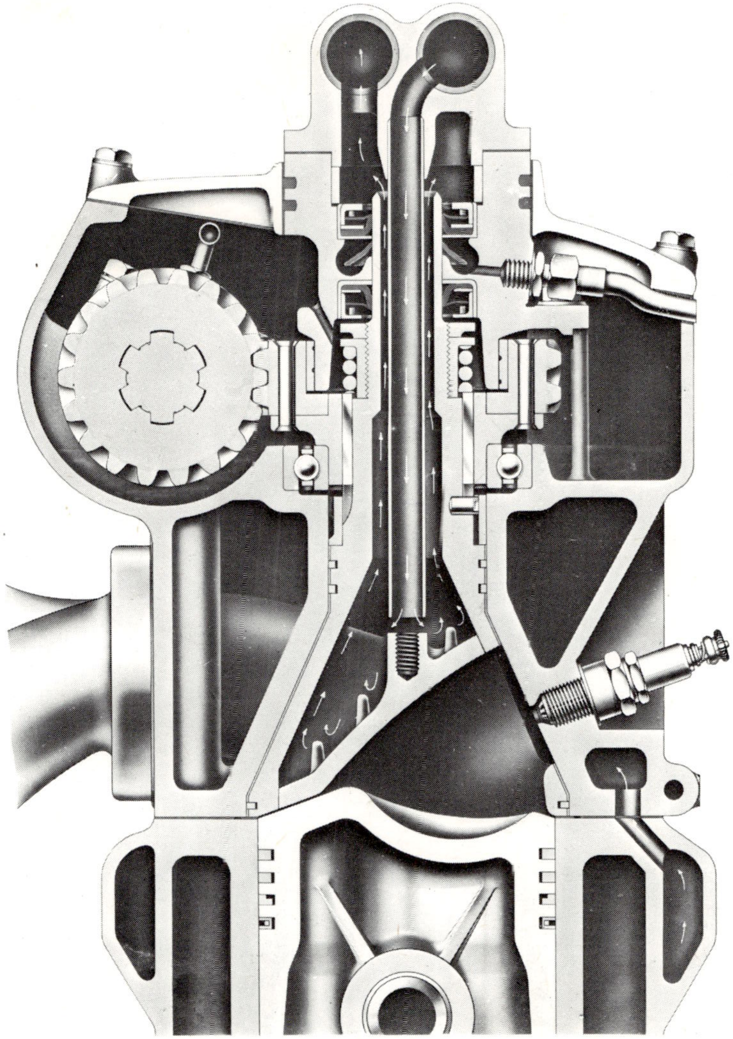

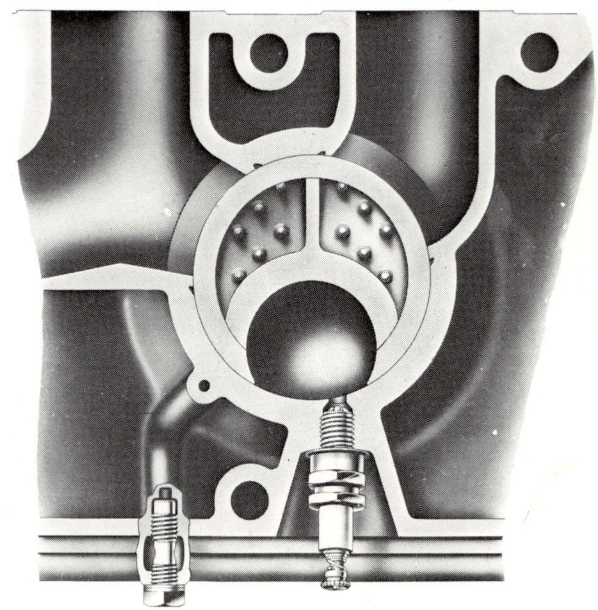

A still later development, and the sparking plug has been transferred to a position at the side of the combustion chamber, whilst the valve has assumed the conical shape typical of all late Aspin models, together with atypical liquid cooling.

Courtesy: F M Aspin.

Lancashire man with a precision engineering business near Bury, where his skilled staff produce some quite astonishing things that it would not be appropriate to mention here. When he started work in 1933 on his rotary valves, he was a teenager; and although the 250 cm^3 motorcycle engine that he adapted ran up to an extraordinary 14 000 rev/min, everybody remained more or less politely sceptical.

Today's Aspin valve is in essentials like the first one, and the development that has been done in the interval provides a fine cautionary tale on the difference between design and development. Was it not Rousseau who first said *Inventer n'est rien*; *construire, c'est peu*; *essayer, c'est tout*? Frank Aspin never gave up trying.

The valve is a conical device, a sort of tap akin in its workings to that time-honoured tongue-twister the brass bung bibcock. Coaxial with the cylinder, the conical valve fills what might generally be considered the combustion chamber. In fact the combustion chamber is a cavity in the wall of the cone, and this cavity is successively aligned with the exhaust and intake ports and spark plug as the valve is mechanically rotated at an appropriate speed. When the piston is at top dead centre, the cavity constitutes the entire combustion space, offering a low surface area in relation to its volume and generally fulfilling most of the requirements for a good combustion chamber shape. It also promises high volumetric efficiency because the inlet and exhaust ports and passages are completely unobstructed when fully opened, because they are opened and closed very cleanly and quickly, and because the duration and area of their opening compare extremely favourably with anything that can be achieved by cam-operated poppet valves.

All these features were satisfactorily demonstrated in the earliest days of the engine's development. Aspin built a number of prototypes, incorporating the same ideas in various forms, and in all cases they proved capable of running at very high rates of revolution, of generating very high brake mean effective pressures over a wide range of engine speeds, at the same time returning uncommonly low specific fuel consumption figures while operating at high compression ratios on low-grade petrol.

There had to be a snag; and in developing these prototypes Aspin found it. The problem was that of maintaining by lubrication a sufficiently low coefficient of friction between the valve cone and its mating surface while controlling the oil so as to avoid an excessive flow during those parts of the operating cycle when the bearing pressure is relatively low. In fact he found that he could build an engine that had reasonably modest oil consumption, or an engine of impeccable reliability, but not both.

In the hope of improving matters Aspin tried detail alterations in the valve

Dismantled twin-cylinder Aspin engine showing drive gear and conical rotary valve cones. Lubrication of the bearing surfaces having always been a problem the 'wavy line' round the 'hem' of the skirt of the valve, which was designed to accelerate the dispersion of a film of oil, will be noted. This version is distinguished by the employment of a cyclically varying gear drive.

Courtesy: F M Aspin.

design. Cones with thin flexible edges, cones with sealing rings, cones located positively in ball bearings, cones floating subject to spring pressure, they were all tried without complete success. Having been trained in metallurgy, Aspin then embarked on a long course of experiments with various pairs of bearing materials, and eventually succeeded in evolving a bearing alloy of his own, and a technique for bonding it to the female surface of the cylinder head. He claims that with this material he can operate the valve satisfactorily without lubrication, but of course the friction horsepower it then absorbs at high speeds is wasteful, while at low speeds the torque that would have to be applied to the valve could in some circumstances be troublesome – and in any case the coefficient of friction is so attractively low when lubricating oil is introduced that it would be folly to try doing without it.

Tribological problems were also approached from another angle. Reasoning that variations in cylinder pressure were responsible for his difficulties in maintaining a controlled oil flow, he sought a means of varying the rubbing speed between the cone and its seat, so that the relative velocity was least when the cylinder pressure was highest. In this way the flow of oil necessary to maintain an unbroken oil film could be reduced by quite a lot. A more complex drive mechanism would be necessary, which would be a shame in something otherwise so simple; but eventually Aspin decided that the end might justify the means, and inserted a simple planetary train above each valve, driving the cone through the planet carriers. In this way (since simplified) the rate of valve rotation could be varied cyclically to slow it right down during the combustion phase (when bearing pressure is highest), speeding up to a rubbing velocity three times higher during the induction phase when pressure is lowest. The side-effects are beneficial: inlet opening and closure is very rapid, stimulating the kinetic flow of the incoming charge and allowing pulse tuning of the inlet tract to be exploited if desired.

In the recent experimental engines in which this variable-ratio drive to the valve has been proven, the ratio has varied between one-eighth and three-eighths so that the mean rate of valve rotation is one quarter that of the crankshaft. In the original Aspin valve the ratio was 1:2, but the variable drive coincided with another important new development: a combustion chamber with duplicated porting, which incidentally allows the relative speed of valve rotation to be halved.

The original single combustion chamber was dèsaxé, of course. Now there are two ports diametrically opposed in the valve body, the result is a symmetrical combusion chamber and a perfectly balanced valve. Not only does this eliminate a potential source of vibration at high engine speeds, it also encourages a better piston shape with a truly conical crown. It follows from the use of twin ports in the valve cone that there must at least be twin inlet passages through the head if desirably large port areas are to be realised. It is not necessary that both of these be fed with a fuel/air mixture: already it has been found sufficient that one convey a rich mixture and the other only air.

This is redolent of the Honda CVCC, but that is irrelevant. Nevertheless, since the earliest days of the Aspin engine, it has been found that the orbiting of the valve cavity has imparted to the mixture not only a degree of turbulence and swirl but also an appreciable measure of stratification, and this is believed to be partly responsible for the engine's ability to run on extremely lean mixtures.

Despite the provision of twin inlet passages it is reckoned that a single exhaust passage will be sufficient for all but those relatively few engines intended for work at especially high ratings, such as in racing. This is just as well, as the profusion of manifolding might otherwise nullify the advantages of compactness that the engine can offer. Thanks to the axial layout of all moving parts, Aspin has been able to devise a form of engine construction exploiting the cylinder head plan in such a way as to permit an engine to be constructed without any bolts or studs other than those associated with big-end or main bearings. By ingenious but simple means perfectly suited to quantity production and even complete automation, he has devised a system of engine assembly that is exceptionally strong, economical of material, has little labour content, and permits a vertical stripping and reassembly sequence such as is necessary in large marine engines and often desirable in cramped car installations.

Nevertheless it is the breathing and burning characteristics of the engine that matter most. It is an interesting feature of the valve geometry that pulse tuning can be applied quite independently to the inlet and exhaust tracts, for it is quite feasible to achieve tremendous flow efficiency without any overlap whatsoever of inlet and exhaust port openings. In fact I have seen curves comparing Aspin's current experimental engines with the racing Cosworth Ford on a valve area/time basis, and saw how easily the Aspin could without any overflap achieve greater nozzle areas than the four-valve Cosworth design which relies on aggressive cams to give high lift with necessarily long duration and therefore plenty of overlap. So although the variable-speed valve makes the Aspin engine a little more cumbersome, it does bring advantages: with a constant drive ratio some overlap would be necessary, though it would still be much less than with poppet valves.

Proof of the value of minimising overlap is evident both on the dynamometer and on the road in cars that have been fitted experimentally with Aspin engines. Very high bmep can be sustained, without any danger of detonation or pre-ignition, down to very low crankshaft rates. Indeed the characteristic shape of the torque curve is such that it is quite feasible to start a car from rest in top gear, or to drive at full throttle up a hill of such severity that the car will eventually stop, finally faltering only at absurdly low engine speed in top gear. Two curious features are associated with this behaviour: one is that ignition advance is either unnecessary or can be limited to one or two degrees before top dead centre; the other is that the lowest specific fuel consumption is achieved at an engine speed somewhat higher than that at which maximum torque is realised. The former peculiarity is clearly connected with the valve timing and the good combustion; the latter is probably due to the increasing swirl and stratification as the valve rotates faster, promoting even better burning.

Certainly the completeness of combustion must now rank as the engine's principal attraction. As in the case of the diesel engine, whose better features the Aspin somewhat emulates, it is possible to ingest greater quantities of air than can be burned, thanks to the amazingly weak mixtures that the Aspin will tolerate. Thus the exhaust is entirely free from carbon monoxide and hydrocarbons, and is very clean and cool. I recall it being reported back in 1938 that it was possible to put one's hand only a couple of inches from the stub exhaust of Aspin's then latest engine (a 1·8 litre light aeroplane engine) without any discomfort. In the event of future legislation becoming so severe that even the latest Aspin should have to be subjected to decontamination measures, it could be adapted to them very easily: all its features (elimination of overlap, good combustion chamber shape, charge stratification and swirl, late ignition et cetera) which permit the engine to run on petrol/air mixtures as lean as 1/23 also contribute to ease of emissions control, oxides of nitrogen included.

Second only in importance is the Aspin's insensitivity to fuel. Because of its immunity from knock, detonation or pre-ignition, the engine can be run at high compression ratio on low-grade fuel, in which the presence or absence of lead can be dismissed as irrelevant. Indeed the engine can be adapted to run on a wide variety of fuels: the 4½-litre truck engine built experimentally in the early post-war years eventually ended up by burning creosote!

For the reasons already given, specific fuel consumption is most abstemious. A level below 0·4 lb/hp h can be expected as a foregone conclusion over most of the engine speed range at full throttle. The best figure is commonly about 0·36 lb/hp h and in some cases I have seen as little as 0·32 lb/hp h, which is as good as anything I know, including diesels and compound-turbo-charged aero engines.

As pointed out earlier, specific fuel consumption and bmep curves do not appear to be interrelated in the same way as is usual in conventional engines. With breathing and burning so effective, high working pressures are only to be expected: even in engines designed for flexibility, 200 lbf/in^2 is a typical level of bmep to be sustained over a wide range of speeds – for example, between 3000 and 6500 rev/min in the case of the last engine I saw assembled and running on the dynamometer. For a racing engine, 250 lbf/in^2 is quite feasible.

In either kind of engine, spark plugs are no problem. Because the plug is shielded except during the transition from compression to combustion phases, it is kept cool and free from oil and combustion deposits. Soft plugs can accordingly be tolerated even when running at high power ratings, without risking any of the dangers generally associated with such a practice. As for sparkless compression-ignition engines, the Aspin lends itself to use as an exhaust valve in pressure-charged two-strokes, encouraging full exploitation of the principles of the uniflow engine. In fact Aspin is really enthusiastic about this, reckoning that the pressure-charged two-stroke is the engine of the future. Hitherto the uniflow engine has been handicapped by its poppet exhaust valves, which cannot be opened early enough or fast enough to give the requisite performance.

Ironically, this is the only respect in which Aspin's simple original ideal – the abolition of reciprocating valve gear – may still be thought to matter. Nevertheless the elimination of the poppet and all its paraphernalia allows the entire valve mechanism to be dismissed as potential obstacles to high rates of crankshaft rotation. The inertia loads in Aspin valve gear are low, balance is perfect, the fatigue life exceptionally long. Note incidentally that because of the very high ratio of valve or port area to piston area possible with this design, it is no longer necessary to suffer extravagantly large bore diameters in order to achieve sufficient valve area: the designer is thus left free to determine his stroke/bore ratio with other factors (such as piston velocity and acceleration, crankshaft length, or engine height) more in mind.

It has all taken a long time, and on the face of it Aspin may be envied his ability to pursue his ideals for so many years. But his engineering business has had to finance all the research and development work on his engines: there have been years when he has been able to do quite a lot, others when work on the engine had to be suspended entirely. When work has been able to proceed, it has embraced engines of all sorts and sizes, from 0·25 to 8 litres. This variety of engine size has established that the Aspin valve is suitable for big cylinders or little ones, in engines running slowly or very fast, in configurations including in-line six- or four-cylinder designs, boxer fours, vertical twins and singles. Now after final adjustments to the twin-port valve to restore the volumetric and thermal efficiency of the original (a little of which was lost when the first twin-ported valve was tried) Aspin is ready with working engines to prove his claims and perhaps persuade someone in the unholy and almighty armies of the engine-industry Establishment that he was right after all.

The work of Roland Cross on his rotary valve engines started in 1920, and only stopped in 1945 when pressures on him to produce aero-engine components had caused the project to lapse. It was as a result of his engine and related works that he was honoured with the Chairmanship of the Automobile Division in 1957. He died in 1970, and it was one of his sons, Michael, who came to the meeting to describe the work of his father.

The principle of the Cross rotary valve design is essentially simple, consisting of a single cylindrical valve, containing a diagonally disposed web, rotating at half engine speed. One end of the valve connects with the induction system and the other with the exhaust. Two ports in the valve connect with a single aperture in the combustion chamber wall, through which are passed both the incoming and outgoing charges, as shown in the illustration. Cross had split the valve housing and had devised an ingenious system of balanced loading to control the forces on the valve, and this proved to be the complete solution to the sealing problems which had beset other rotating valves. Valve lubrication had also received special consideration and an effective oil recovery system was devised. Thus any wear was automatically compensated and the possibility of seizure eliminated.

Cross had produced a design of symmetrical uncompromised combustion chamber of high volumetric efficiency, in which surface temperatures were controllable to a low order. The exhaust poppet valve, which in a conventional engine may run at a temperature sufficiently incandescent to ignite the incoming charge unless additives were used in the fuel, had been eliminated. The Cross design allowed good uncluttered port areas not limited by bore diameters, and surfaces of the combustion chamber were subjected to approximately equal mass applications of cool and hot charge applied symmetrically about the cylinder axis. This meant that the chamber structure was not subject to concentrated and asymmetrical heat fluxes in critical areas. The rotation of the valve presented a fresh and moving surface to the charge throughout the firing stroke, and the low temperature incoming charge provided the valve with a significant cooling effect. This allowed the use of a high compression ratio to increase thermal efficiency and improve fuel consumption. The theoretical advantages to be derived from the use of rotary valves were amply realised in fact, and as an example a 350cm^3 air cooled engine with a compression ratio of 10·5 : 1 was capable of sustained

Courtesy: M. E Cross.

The rotary valve is not a new idea. The late Roland Cross commenced his development work in 1920, the J & G McGee version made in Tollcross, Glasgow was even earlier, and both Darracq and Itala, among others, enjoyed unsuccessful courtships with the rotary valve principle. The 1938 350cc air-cooled motor-cycle unit shown here employed Cross-designed rotary valves and achieved some gratifying performances.

operations at high loads. A brake mean effective pressure of 195 lb/in^2 was obtained at 4400 rev/min on a 66 octane fuel, with a specific fuel consumption of 0·39 lb/bhph. The engine is illustrated in the photograph.

Bearing in mind that these results were obtained over 35 years ago on a relatively undeveloped engine with a simple non-optimised Amal carburettor, these results are truly outstanding. Cross was never reluctant to subject his engines to examination, and results obtained by independent test houses amply confirmed his findings. Cross built many rotary valve engines, both single- and multi-cylinder, and all enjoyed the same characteristics of flexibility, smooth power curves and flat torque curves. He was responsible for some other interesting engine developments, including the unlinered aluminium engine and the steel piston ring for gasoline and diesel applications. He also established the company which bears his name.

While I am not able to disclose details, I have heard that a new 4 cylinder engine is currently under development in England, containing Cross rotary valves and the Baker crank mechanism. I understand that the objective is to produce a light and compact installation using existing technology and conventional materials. In view of the emission regulations and acute fuel shortages, it is hoped that the proven ability of the Cross combustion chamber to accept low octane fuels at high compression ratios with low specific consumptions will provide a decisive advantage.

However, the general case of the rotary valve is not lightly to be dismissed. For presenting the maximum unobstructive throughway for the longest period there can be little to rival it, though much will depend on choice of port dimensions. Opening and closing angles are certainly no criteria, for the rotary's valve moves with a constancy of rotational speed that is quite different from the acceleration and deceleration of a cam-operated poppet valve – even in the case of Aspin's variable-speed rotor. However, because the timing of the rotary valve is so consistent (that is, its angular velocity is substantially constant) the theoretical length of the opening period is less than that for a poppet valve offering an equivalent maximum throughway. What is more, it seems to be a common feature of rotary valve engines that the bmep curve and all those related to it are uncommonly smooth, without any of the peaks and lacunae to be observed in the curves of poppet-valve or other types. The working range of the engine also appears to be extendable to both extremes – at low speeds because of the better timing of port opening and closure, at high speeds because of the absence of mechanical limitations due to reciprocating inertia loading. From the commercial point of view, the rotary valve has the advantage of being adaptable to existing engine designs since the power unit is not significantly altered below the cylinder head

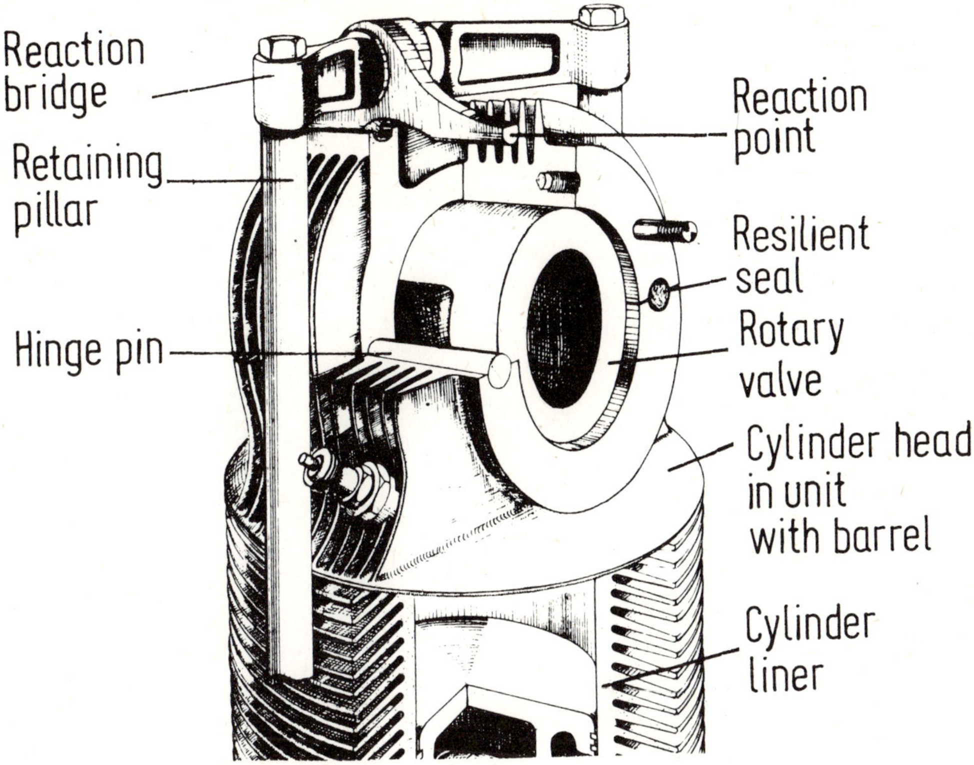

Sectioned view of the 1936 Cross rotary valve air-cooled motorcycle engine adapted from a 250cc Rudge unit. It incorporated controlled valve loading.

Courtesy: M E Cross.

joint level. Nor do the components and their drives require anything difficult or novel in the way of production techniques. Only thermal and tribological problems seem to be present, and it is tempting to think that the rising generation of ceramics and cermets (notably such extraordinary materials as silicon nitride, which could easily be pressed into the form of an Aspin rotor and might prove more suitable for the duty than its low heat transfer would suggest) may be the salvation of the idea.

The sleeve valve is at first sight an obvious rival to the rotarics, but its virtues are most applicable to pressure-charged engines. The sleeve valve is nowadays too little remembered, and in its time was too little respected, though its workings are already well enough documented for an extensive review to be unnecessary in these pages. There are only three firms that really made it work as well as it might, though numerous others dabbled with it in the 1920s: Napier were most successful in getting the sleeve valve to work really well in their prodigious Sabre, and Ricardo must be acknowledged as providing the original impetus and encouragement for sleeve valve development, while it was Bristol who really put sleeves on the map. It is worth pointing out that sleeve valves featured in the most powerful piston engine to fly (the 5500 hp version of the Napier Sabre) and in the most reliable (the last version of the Bristol Centaurus, which holds the record for piston engine overhaul life at 3000 hours), and have been proven to allow higher compression ratios, higher exhaust back pressures, and higher outputs, for a given engine capacity or weight than poppet valves could. The early difficulties experienced with sleeve valves are too well remembered, and engineers recalling them are too ready to accept as dogma the notion that sleeves might be serviceable in slowly and steadily-turning aero engines but could not be adapted to anything sprightlier. All that can be said in argument may be summed up by the facts that Bristol ran one of their 6½ inch sleeves up to 8000 rev/min in a single-cylinder test engine without anything going awry, and that the Napier Sabre could blip up and down the rev-counter scale like a car engine.

As at various times in the past, experiments are still going on in the hope of procuring with poppet valves some of the performance characteristics that have already been recited to the credit of rotary valves and, to a lesser extent, of sleeves. Variable valve lift or duration of opening has always been a popular ambition of engine designers, as has mechanical operation of valves without reliance on springs. Mysteriously, very little effort has been made to develop other kinds of springs than the conventional helical variety, though we have seen a few torsion bars in such engines as the Panhard Tigre and the Honda CB 450. As for positive mechanical closure of valves without the

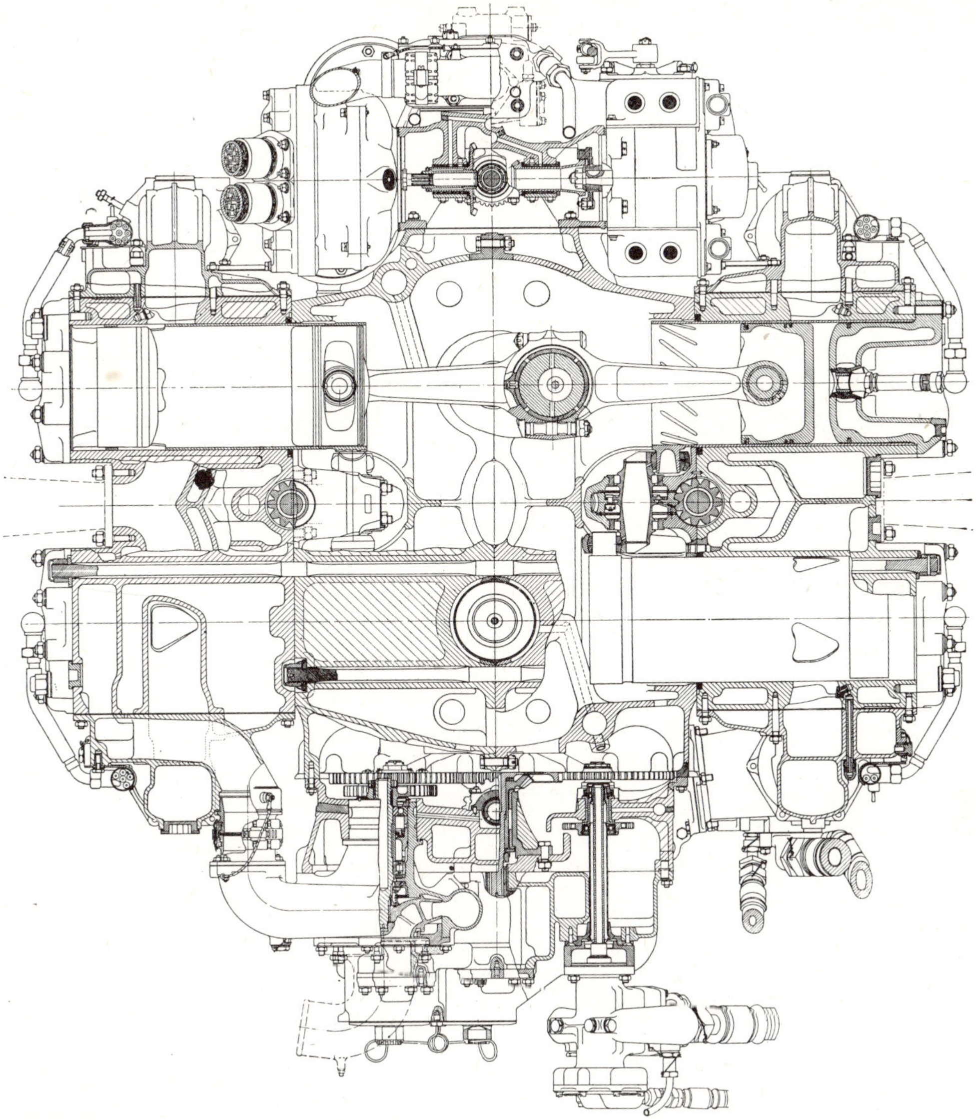

The British policy of combining high boost pressures and high speeds in aero engine design was carried furthest in the Halford-designed Sabre, the geometrical arrangement of which was a Napier fashion that broke a pattern of radial and V-12 layouts which had dominated the scene for 20 years.

Liquid cooled, and with 24 cylinders, the Sabre was basically two 12-cylinder horizontally-opposed engines superimposed one upon the other, but sharing a common crankcase. The compactness thus achieved was supported by the use of Burt-McCollum sleeve valves, and in this rigid form and displaying a very small frontal area, the engine successfully powered the *Typhoon*, *Tempest* and *Fury* (the last at well over 480 mph at 20 000 feet) during and after World War 2.

use of springs, popularly known as desmodromic operation, the idea is of considerable antiquity and designers have returned to it from time to time for the engines of racing cars and motor cycles. Some of the earliest efforts were quite successful, as in the 1911 Grand Prix Delage, but it was not until the Grand Prix car had grown up into the 1954 Mercedes-Benz that desmodromic operation was perfected to the extent that no springs whatever were employed.

The construction of the Mercedes-Benz valve gear was straightforward enough, although as might be expected the materials and workmanship were of the highest quality: but the interesting and then unique feature was the elimination of the little final closing spring that had always been thought necessary. Daimler-Benz found that by allowing a trivial amount of backlash in the mechanism, and contenting themselves with positive cam return of the valve to within 0·3 mm of its seated position, the valve's momentum would ensure that it completed the journey, and gas pressure in the cylinders would do all that was necessary to keep it there.

For a little while before the brief memory of the modern motor racing enthusiast expunged all recollection of the cars, it was fashionable to cast aspersions on the Grand Prix Mercedes-Benz machines of 1954 and 1955, in terms suggesting that they were not as fast nor as powerful as the expensive and idiosyncratic design and construction ought to have made them. This is an unfair judgment, made because Daimler-Benz suddenly and unforeseeably decided to give up racing during the course of the 1955 season. It is known that the car was originally intended to have had a longer competitive life, and that it was designed with a view to progressive development over perhaps twice as many years as it enjoyed. There is reason to suppose that by the end of 1957 the cars would have boasted a considerably higher performance than that which proved sufficient to carry them successfully through the 1954 and 1955 seasons.

During the course of those two years of competition, the power output of the engine was increased by more than 13 per cent from 256 bhp at 8250 rev/min at the beginning of 1954 to 290 at 8500 at the end of 1955. At these levels the desmodromic valve gear and the colossal valve openings that it made possible were not being exploited other than as a safety feature to improve mechanical reliability; but there seems little doubt that had full advantage been taken of these mechanical features (and of others such as the variable-length inlet tracts which were proved experimentally to fill out the torque curve in a most attractive way) the cars could have reached the 1957 season with bmep levels at least similar to the best of their opponents – which would have meant far more power than any of them, to the tune of at

Worthington-Williams Collection.

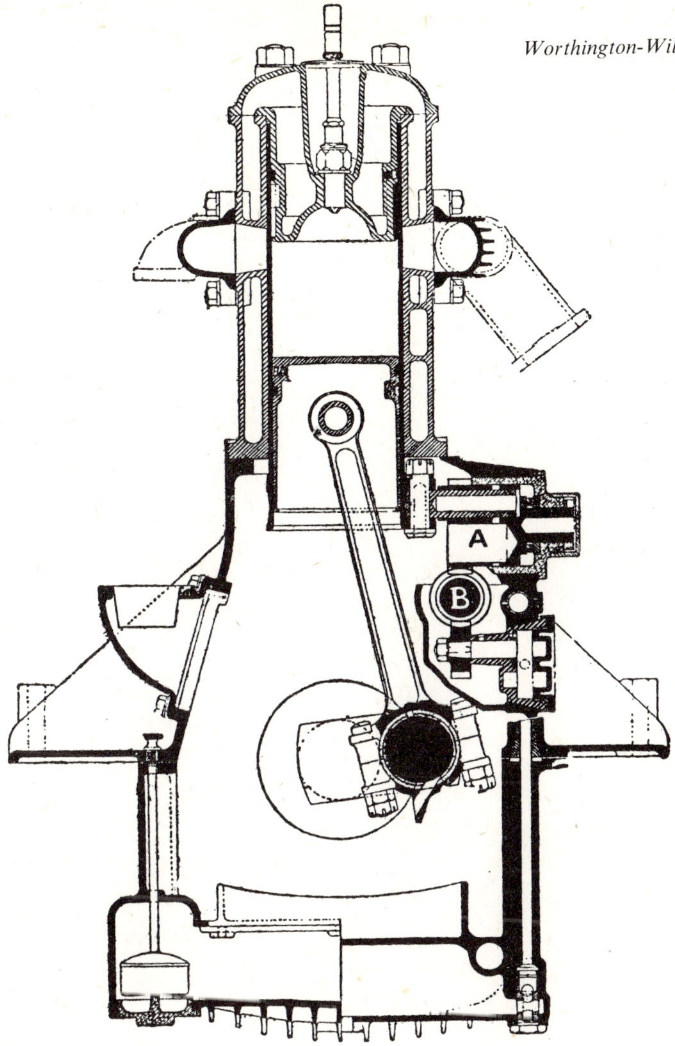

Phoenix revived. It is ironic that a basically sound design of single-sleeve engine—which the Burt-McCollum most certainly was—should have forced the mighty Argyll Company to its knees, and the marble halls of Alexandria into the hands of their Lordships of the Admiralty (it became the Naval Torpedo Factory) only 11 days before the events in Sarajevo in June 1914 which precipitated World War 1. The irony is compounded when one considers that the financial collapse had been hastened not only by expensive litigation with the Charles Yale Knight interests, but also by development expenditure on a single-sleeve-valve aero engine which had already received the blessing of the Royal Aircraft Establishment. After passing into the hands of Continental Motors Corporation in 1925 (who did nothing with them) the Burt-McCollum patents were finally acquired by the Bristol Aeroplane Company in 1934. Under the direction of the late A H R (later Sir Roy) Fedden the Perseus radial aero engine was introduced and marked the first of a line of Burt-McCollum types which played a dominant role in the 1939-45 conflict in *Wellingtons*, *Tempests* and other aircraft. Argylls were always ahead of their time!

1954 Mercedes-Benz 2½ litre M196 GP engine. With direct injection and desmodromic valve gear this eight-cylinder unit was instrumental in winning Mercedes-Benz the Manufacturers Championship and Fangio the World Championship in 1954 and 1955.

Courtesy: Daimler-Benz Aktiengesellschaft. Stuttgart.

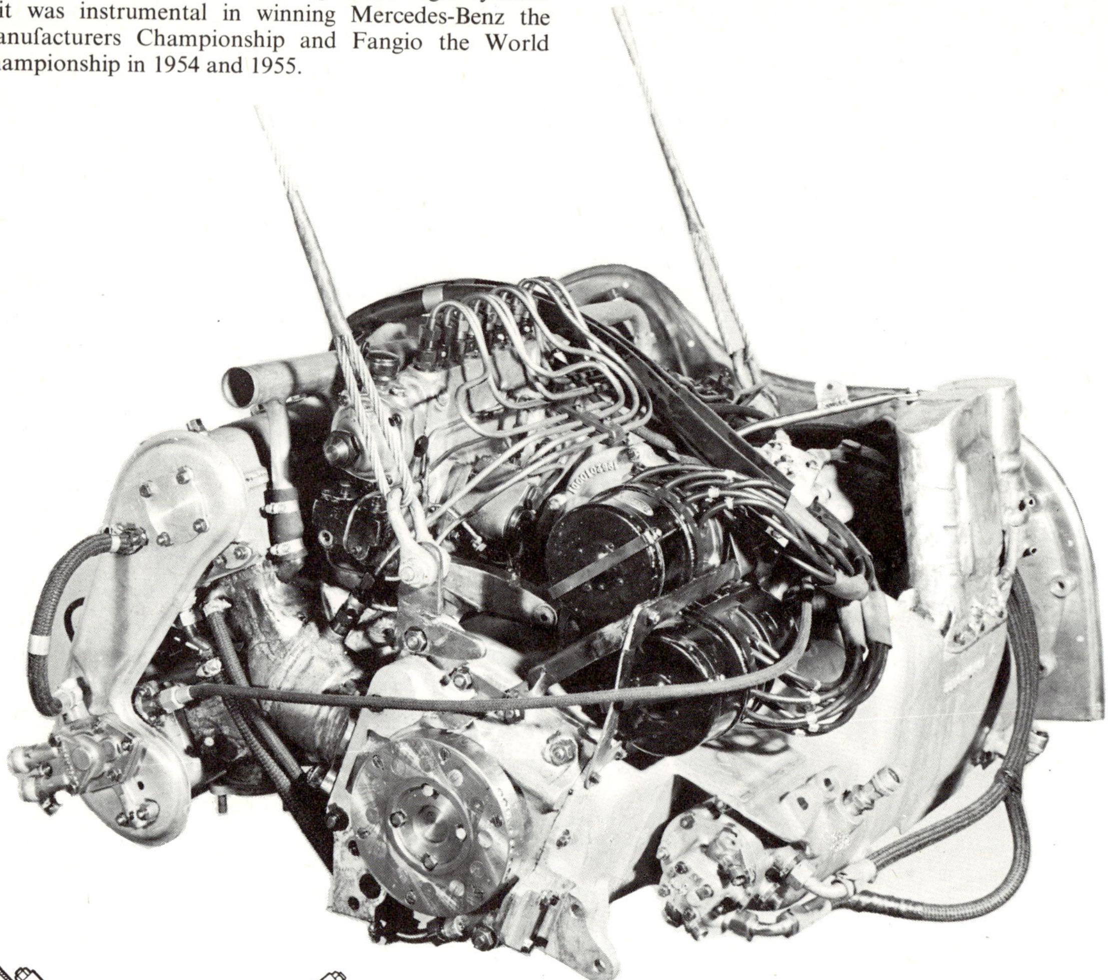

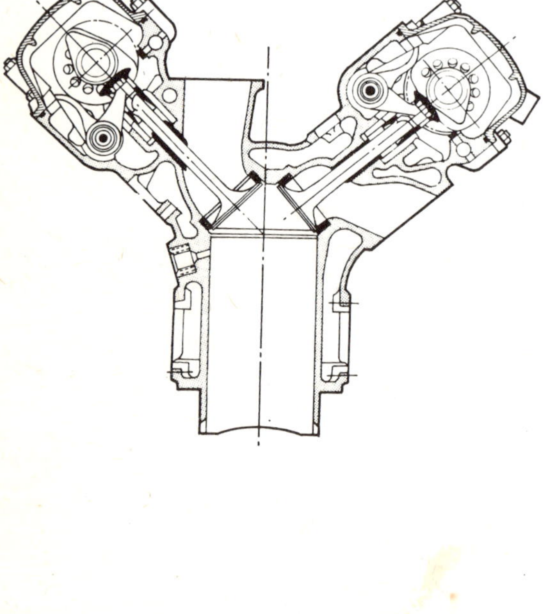

Desmodromic valve gear of the 1955 Mercedes-Benz M196/110, a three-litre sports car version of the M196 GP engine. It was highly successful in competition, the 300SLR winning most of the major sports car events in 1955, including the Mille Miglia.

Courtesy: Automotive Design Engineering.

least 340 bhp without any extension of the rotational gamut beyond 8500 rev/min and perhaps the target 390 bhp at 10 000 rev/min for which the valve gear was designed.

Instead, the whole thing was shelved; and the only manufacturer thereafter to exploit desmodromic valve operation effectively was the Italian motorcycle firm of Ducati, whose machines were virtually unbeatable until after many years they became obsolete in other respects.

Running variation of poppet valve lift and event need not rely on desmodromic operation for its success. The most exquisitely refined version yet seen allows the valves to be returned to their seats by springs in the usual manner, but the operating mechanism does everything to ensure that the engine should combine the best features of the most rabid and the most restricted of conventional types. This is the experimental Fiat engine conceived and developed by engineer Torazza.

The essence of the Fiat Torazza engine comes from early developments announced in 1970 by General Motors who had been experimenting with intake valve throttling. Here, the pivoting rocker levers of an otherwise fairly conventional pushrod engine were given movable fulcra enabling the intake valve lift to be varied. The object was to correct the cyclic variations of cylinder pressure suffered by a conventional engine digesting very weak mixtures. By minimising the valve lift so as to create the smallest possible opening for the load applied, and thus maintain the full pressure drop across the port, the velocity of the incoming charge is kept up to the speed of sound. The droplets of petrol passing through the port undergo almost explosive atomisation in these circumstances, resulting in fast and consistent combustion immediately after ignition, even when the engine is running lean at part load.

The object, as might have been guessed from the date of the announcement, was to reduce toxic emissions in the exhaust. Torazza took the idea further, arguing that the possibility of changing both intake and exhaust valve opening duration and lift as a function of speed and load must be even more attractive. The essential mechanical features were an oscillating cam driven by an eccentric and connecting rod, and finger followers bearing on the valves. By moving the fulcrum of the follower in an arc swung from the cam's centre of oscillation, timing and lift were altered simultaneously – lower lift and softer timing in one direction, higher lift and sportier timing in the other. The typical valve opening event was varied between extremes of 224 and 294°, the corresponding variation in lift amounting to 37 per cent. Control was effected hydraulically through a three-dimensional cam moved axially in response to engine speed and radially to inlet manifold depression.

The snag encountered by General Motors with their intake valve throttling was that in light-load running, when valve lift was very small, despite the high airflow velocity across the seat there was not enough movement upstream in the induction tract to support the droplets of fuel sprayed by the carburettor jet or conventional injector nozzles. This objection could be overcome by the truly atomised mist produced by the electro-pneumatic apparatus more recently developed by Petrol Injection Limited as a more acceptable successor to their earlier Tecalemit-Jackson equipment. However, engineer Torazza has not reported any such difficulties, on the contrary finding no increase in the engine's tendency to detonation despite evidence of a leaning out of the mixture. This he though might be due to an improved mixing of fuel and air downstream from the carburettor, as well as to the higher gas velocities across the valve opening annulus.

The experimental results verified by Fiat for the Torazza engine are wholly encouraging. An increase in bmep of over 30 per cent is realised at low speeds, drooping to 5 per cent at high engine speed where the size of the inlet and exhaust passages is the controlling factor. We should bear in mind that the conventional engine with which it was compared was camshafted for optimum performance in a regime between 60 per cent and 90 per cent of maximum power speed. It is worth mentioning that the whole arrangement was simply superimposed on normally-produced four-cylinder Fiat car engines, in one case with inclined opposed valves in the standard hemispherical combustion chamber, and in the other with in-line valves in the standard wedge-type combustion chamber.

Other points to note were that the specific fuel consumption was improved and that valve gear noise was no greater than in the standard production engines. No abnormal vibrational fatigue was observed in the valve springs or the kinematic system, nor any abnormal wear in any engine component after 600 hours testing, much of which was at maximum power and at overspeed.

Inevitably such mechanism is more costly than the normal arrangement because a greater number of components demand more production operations and more complex assembly. A study of the torque characteristics of the engine make it clear, however, that if the unit is designed *ab initio* for this variable timing system, the increased cost may be compensated by the better performance and the possibility of reducing the engine displacement or, more probably, the number of transmission ratios and thus the overall weight of the power train. There are more possibilities inherent in the system, notably that of obtaining better mixture homogeneity through valve throttling at part-load and an internal exhaust gas recirculation by means of

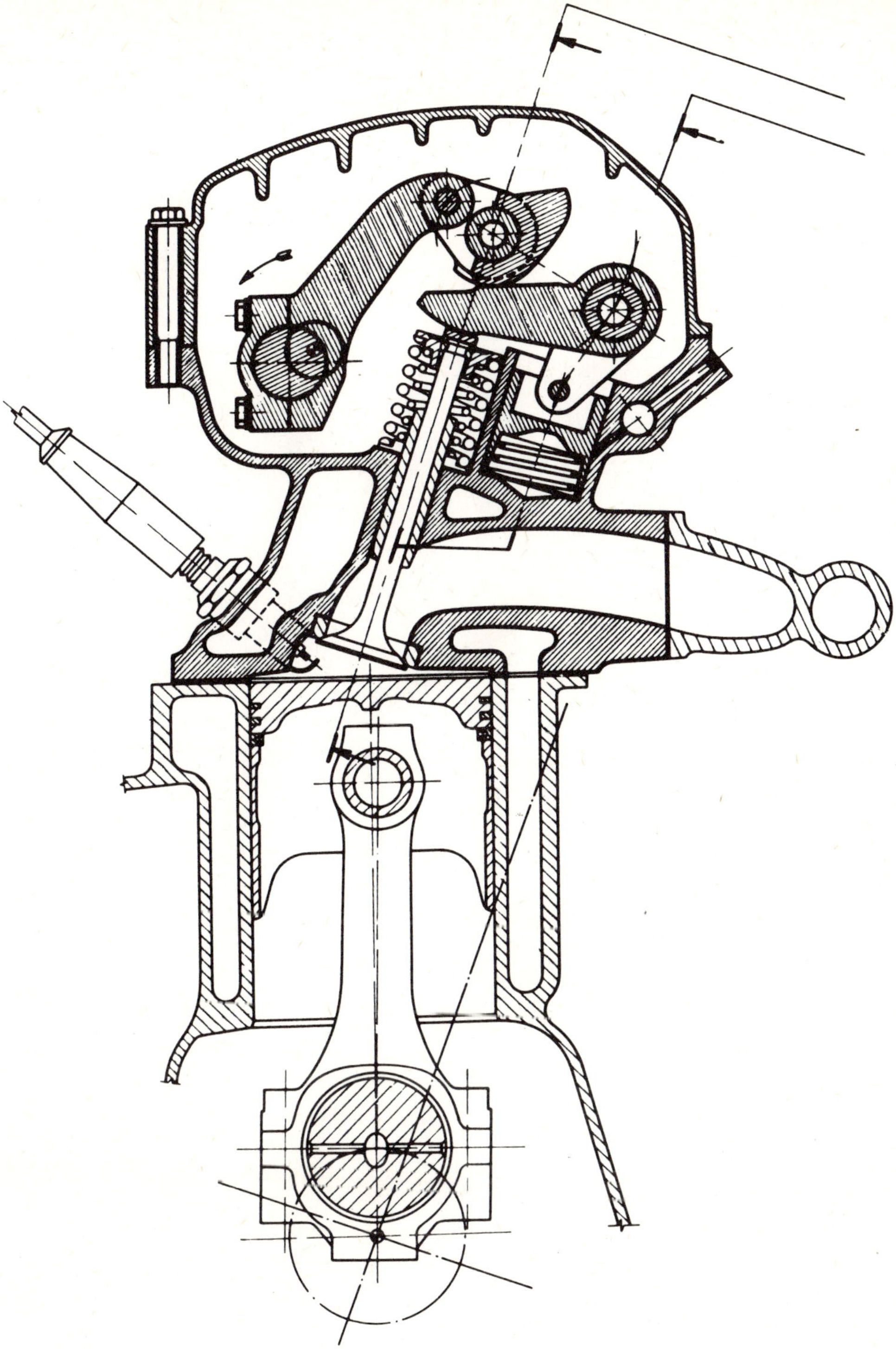

The Torazza-designed scheme of valve drive by a combination of oscillating cam and rocker, developed by Fiat in 1971.

External view of the Fiat Torazza engine, partially disassembled. Valve mechanism shown inset.

negative valve overlap. Moreover, some of the characteristics offered by this variable valve-gear may be of value not only in competition engines but also in the diesel and multifuel field: chief of the attractions is the possibility of increasing the engine's breathing capacity and turbulence throughout its speed and load range, and the possible improvements when starting at cranking speed.

Indeed there is only one objection to the Torazza Fiat engine in the context of this book. When you look at it, as I did early in 1971, you see that it does not look at all unusual.

4. Engines of Unusual Format

It was once stated by a certain gold-digging lady that beauty was more important than brain because the average sugar daddy could see more clearly than he could think. The principle is one that might well be borne in mind by the engine designer: should he seek to win proselytes by expounding the virtues of some heterodoxy of combustion or ignition or working cycle, most of his customers and probably his principals too will soon be lost, whereas if he should propose an engine that looks peculiar but works conventionally there is ample historical precedent to suggest that they may welcome the design. So long as it has perfectly ordinary pistons and so forth inside it, people seem quite eager to accept an engine that has a funny shape.

In the first twenty years of this century, the greatest variety of such oddities were to be found among aero engines. The first good ones – the exquisite radial of Charles Manly, and the fuel-injected in-line engine of the Wright brothers – were in retrospect conventional enough; but they were soon followed by a host of oddities, all enthusiastically propounded with more thought for the problems of flight than for the niceties of engine design. Thus REP (otherwise Robert Esnault-Pelterie) built fan-type engines which, though otherwise conventional, featured cylinders stuck on the crankcase at all angles, five, seven or even ten of them to produce a lot of vibration and anything from 25 to 50 hp. Even odder was the Faccioli engine built by the SPA car firm: this engine employed the opposed-piston layout already seen in cars (notably the Gobron-Brillié), but Faccioli took the idea further by having two banks of superimposed cylinders which were neither coaxial nor parallel (save in the vertical plane); the engine was in effect two narrow-angle V twins or V4s opposed so as to have common combustion chambers between opposing cylinders and a horizontal-axis crankshaft at each end, the idea being to couple an airscrew direct to each shaft. These weird devices were built in a variety of sizes and powers, ranging from a 20 hp 2-piston version to a 100 hp 8-piston affair. All were water-cooled, some were surprisingly light, some were intended for dirigibles, and none enjoyed the slightest

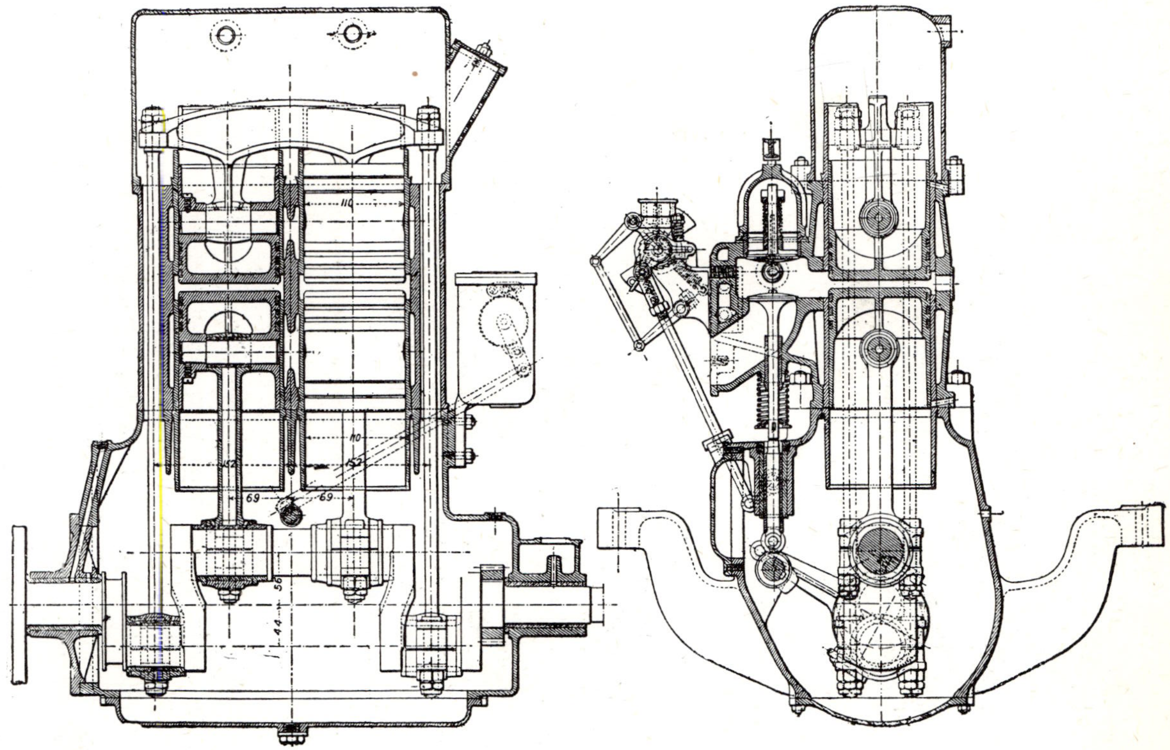

'Gobbling Billy'. The unusual exhaust note of the opposed-piston Gobron-Brillie, not to mention its somewhat irreverent nick-name, did not deter prospective purchasers, and the design remained in production in private and commercial form for over 20 years, finally giving way to conventional types in 1922. Normally offered in two-cylinder four-piston form (although there was a terrifying six-cylinder model, the 70–90 hp of 1908), the lower pistons coupled direct to a common crankpin by normal connecting rods, the upper ones relying upon a cross-head connecting them via twin tubular connecting rods to the lower crankpin, and actuating crank throws opposed at 180 degrees to those of the rods of the lower pistons. A freedom from vibration was claimed by the makers which was probably more comparative than absolute.

Worthington-Williams Collection.

popularity – so by 1912 Faccioli had given up his nonconformist ideas and turned to the design of conventional water-cooled V8s and air-cooled radials. These proved of no use either.

If any proof were needed that morphological oddity was not a bar to success in the aero engine business, the brief but convincing domination of the business by the rotary engine should provide it. The principle had been worked out as early as 1887 by an Australian, Lawrence Hargreave, for a compressed-air engine; but it was the mechanical ability and acumen of the French brothers Séguin that really put the rotary on the map. The popular Gnome, with one of its valves in the centre of the piston crown, was undoubtedly peculiar; but in 1913 the firm produced a new version in which the inlet valve was replaced by a system of piston-controlled ports, and this *monosoupape* or single-valve engine promised to be less troublesome than the two-valve original. It was certainly very ingenious: air was drawn into the crank chamber from the front of the engine through the open end of the hollow nose-piece and crankshaft, emerging from which it was drenched with petrol so that in normal running conditions the mixture in the crankcase was too rich to support combustion. Now suppose the piston to be performing its downward working stroke: the exhaust valve is opened early, allowing the combustion gases to escape to atmosphere so that when the piston uncovers the ports low down in the cylinder wall (communicating with the crankcase interior) little or no flow of gas takes place in either direction through them. On the succeeding upstroke, the exhaust valve remains open and the residual exhaust gas is pumped out. Then the piston starts moving down again on the suction stroke; but instead of the exhaust valve closing, it is held open until about one-third of this stroke is performed, thus permitting fresh air to enter the cylinder. When the exhaust valve is closed and a partial vacuum is created within the cylinder by the continued downward motion of the piston, the cylinder ports are again uncovered and in rushes the rich mixture from crankcase to be diluted with the fresh air already in the cylinder and to be compressed for the next combustion cycle.

Odd it may have been, but it worked. The pilot was able through simple linkages to vary the extent and duration of opening of the exhaust valves, a method of speed regulation so effective that these engines (whose rated speed was 1200 rev/min) could run at as little as 200 rev/min. On the other hand, this method of regulation caused rapid burning of the exhaust valves.

There were plenty of other rotaries even more outlandish in design, for the early success of the Gnome encouraged inventors to extravagances of ingenuity. There was for example the six-cylinder 60 hp EJC rotary, in which cylinders and crankshaft rotated in opposite directions. Each might have

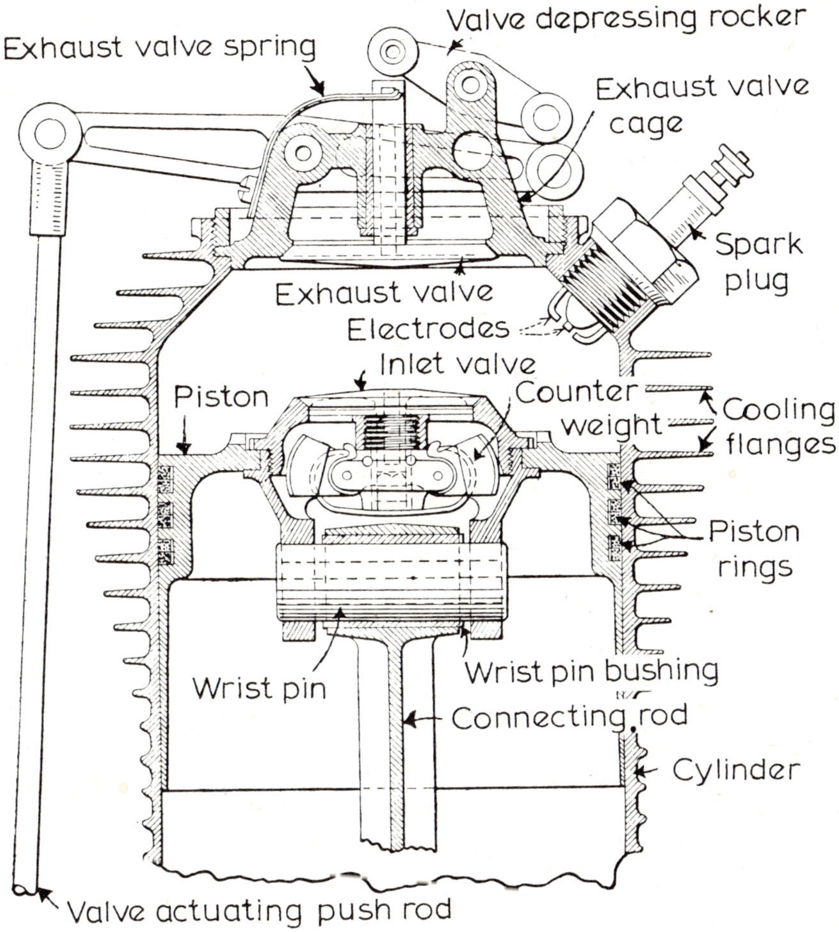

The 7-cylinder Gnome rotary, designed by Laurent Seguin in 1907, was an elfin thing with cylinder barrels of only 1·5 mm wall thickness, albeit machined out of hefty nickel-chrome steel ingots, fins and all. The single valve apparent in the head might suggest a uniflow two-stroke, but with an odd number of cylinders the engine clearly ought to be a four-stroke. The secret was that the automatic inlet valve was concealed in the crown of the piston, where it communicated with the carburettor through the crankcase. In 1913 the metallurgical hazards implicit in this arrangement were avoided by the design of the Gnome Monosoupape rotary, which really did have only one valve per cylinder and was incidentally much more manageable—and despite its piston-controlled transfer ports it was *still* a four-stroke.

Courtesy: Science Museum, London.

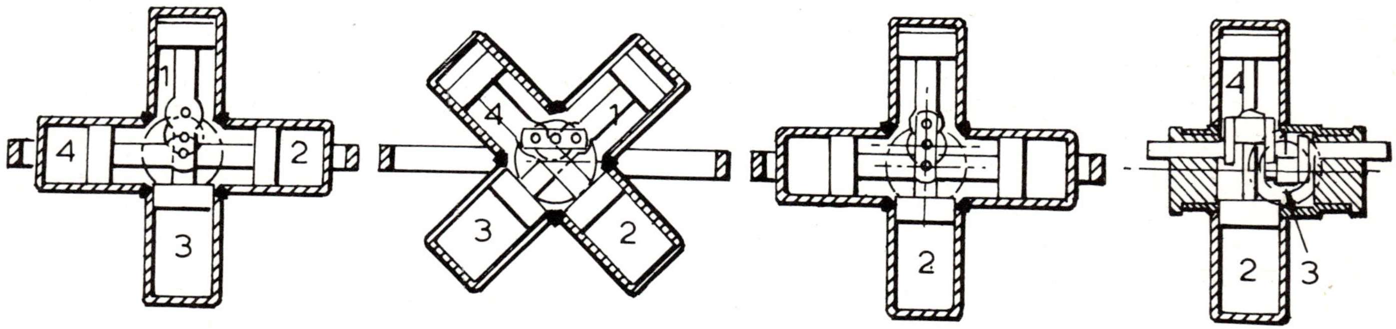

Moteur rotatif: Lyon was once something of a centre for the motor industry in France, and over a long period spawned a great many designs—some more successful than others. Antoine Burlat commenced his experiments with rotary engines in 1904 and applied these to motor cars in 1905. They were unsuccessful, however, and after a period during which C A R (Camions et Autobus à Moteur Rotatif) lorries were built, Burlat turned his attention to aero engines, still employing the rotary principle, and between 1911 and 1913 examples were flown in Dufaux and Voisin biplanes.

Geometrically very similar to the Karol-Ansdale engine, the Burlat rotaries suffered from the limitations inherent in their type, and by 1919 the company were concentrating their efforts on a propeller-driven bicycle!

Courtesy: L Loreille.

an airscrew attached, thus antedating by twenty years the double-twelve AS6 produced by Fiat for the Schneider Trophy and used for capturing the air speed record. The idea might not have worked so well with the EJC, for the difference in speed between the two rotating systems was of the order of 3:2, and airscrew design was hardly ready for this sort of thing. Siemens and Halske built contra-rotaries in which both elements rotated at equal speed, but fared little better. Then there was the Burlat rotary, which made use of the straight-line motion produced by hypocycloid gearing such as had been invented by a Mr Abraham in a four-cylinder combustion engine of 1898, following its invention by Sir Charles Parsons in connection with a steam engine in 1877, prior to which it had been much employed in textiles machinery. However, the Burlat engine was never successful; and neither was the double-acting six-cylinder Demont rotary, nor the geared rotary two-strokes by Laviator, Lamplough and others.

Gradually, however, opinion hardened in favour of the fixed cylinders and revolving crankshaft that we all know and accept today. It was a layout more supported by convenience than by authority, having been patented and developed for steam engines by one James Pickard, a Birmingham button maker, in 1780.

As the rotary engine fell prey to its own limitations towards the end of the Great War, ingenuity in the layout of fixed-cylinder engines became manifest on all sides. Undoubtedly the most outstanding such layout (though too late for service before the war ended in 1918) was the Napier Lion, a 12-cylinder engine laid out in three banks of four cylinders to make a broad arrow of 60° included angles. The cylinders were all separate, but the heads were monobloc castings screwed to the flat-topped ends of the cylinders by threaded valve seat inserts. It was to prove a superb engine, and its designer (the gifted A J Rowledge, who later went to Rolls-Royce and revitalised their design thinking) had the satisfaction of seeing his brainchild remain in service for decades afterwards, establishing an enviable reputation for power, economy, reliability and versatility.

However, Napier did allow themselves to be briefly sidetracked just after the end of the war, producing a curious 16-cylinder device called the Cub. This engine consisted of four rows each of four cylinders arranged in an irregular cross around a common crankcase, with the angle between each upper cylinder block and its corresponding lower block being 90°, the angle between the two upper blocks being 45°, leaving 135° between the two lower blocks. Improbable though it looked, this was an excellent configuration when combined with a single-plane four-throw crankshaft, giving firing intervals of 45° and perfect primary and secondary balance – whereas the

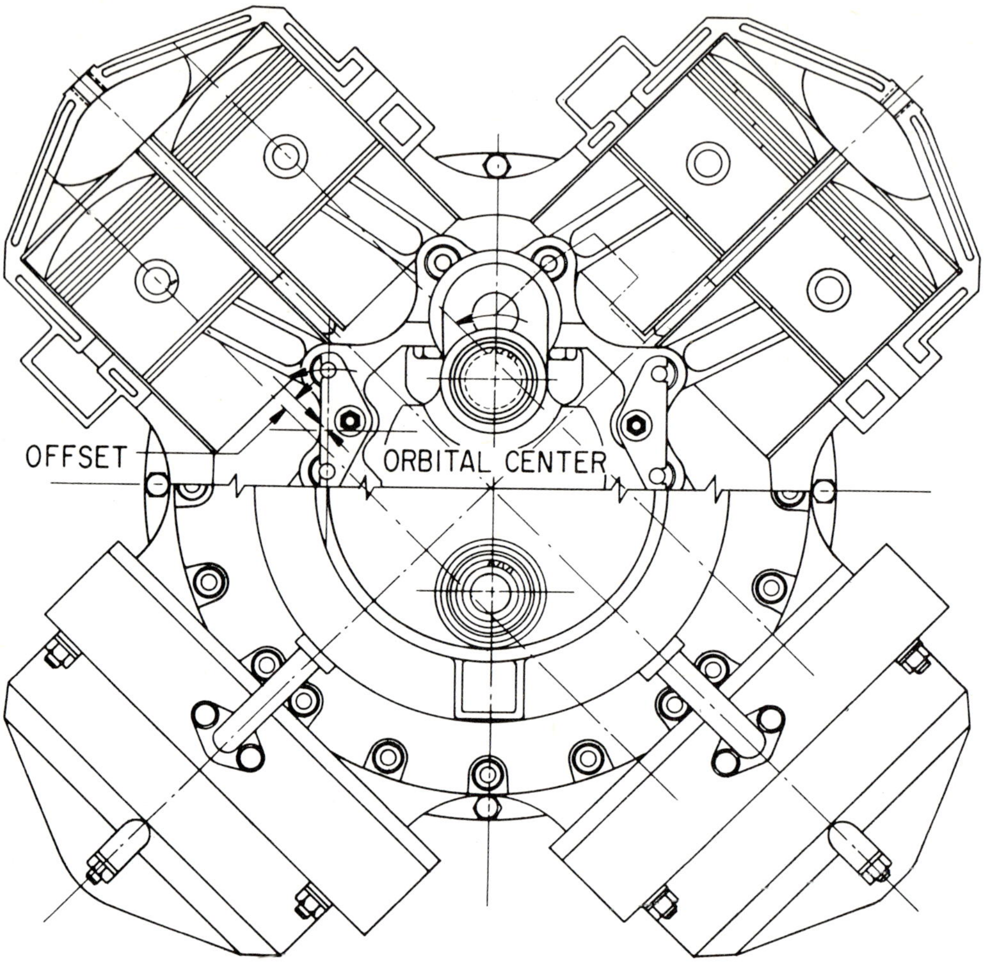

The Thiokol Dyna-Star engine developed by the Thiokol Chemical Corporation of Pennsylvania in the early '60s incorporated a number of interesting features, some of which were entirely new. Working in four pairs of cylinders arranged in cross formation on two radii mutually at 90 degrees, the connecting rods did not link the pistons directly to the two crankshafts; instead, they were coupled to an octagonal frame carried on the crankpins. A two-stroke, with non-detachable heads, and a common combustion chamber to each pair of pistons (in which respect it was similar to the Duplex "four-eight" of 1920) it was intended for use as either a compression-ignition or spark-ignition unit of compact dimensions. By altering the geometry of the frame and the relative inclination of the cylinders it was possible to vary compression ratios and timing, and to exploit *désaxé* to reduce possible side thrust. Unfortunately development was discontinued circa 1969.

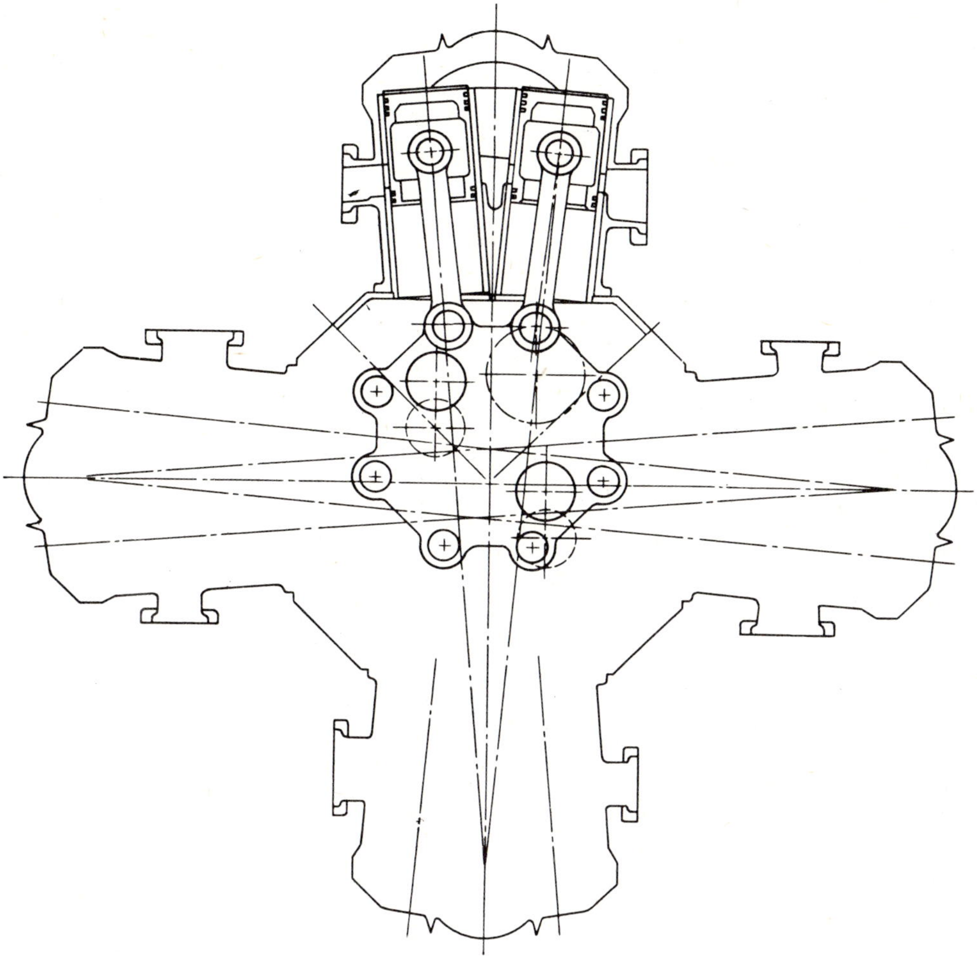

broad-arrow layout of the 12-cylinder Lion was left with a slight horizontal secondary shake. The Cub was nominally a 1000 hp engine, but it developed 1057 hp during its first run on the test bed, which was practically a running-in session and certainly not a tuned-up run. At the time it was in fact the most powerful aero engine in the world, though in France a broad-arrow 24-cylinder Lorraine Dietrich soon rivalled it while in Germany Rumpler designed (but was not allowed to build) a 1000 hp 28-cylinder seven-row radial engine, each bank of cylinders having an array of concentric rocker shafts transmitting motion to the valves from cam rings at the rear of the engine.

Many other curious configurations were dreamed up at about this time. They included engines of barrel type, or with horizontally opposed cylinders, or with vertically opposed pistons, or even that extraordinary monstrosity the Breguet-Bugatti which consisted of two Bugatti engines set in line and turned end-for-end, with the crankshafts of one at a lower level than those of the other and all four shafts geared into a common spur drive. The Bugatti aero engine itself was peculiar, consisting essentially of two vertical straight-eights sharing a common crankcase, with their two crankshafts each geared into a common spur wheel driving the airscrew shaft. In fact any Bugatti engine at the time was entitled to be considered an oddity, but they are all sufficiently well documented to make further description unnecessary here.

However, Rumpler's multi-bank quest for power found its echo in America at the end of the next big war. Lycoming embarked on a colossal engine, rated initially at 5000 hp and intended ultimately to achieve 7000 hp. This monstrous folly was a liquid-cooled four-row radial with thirty-six cylinders to give a total displacement of 127 litres: the engine was over ten feet long and weighed 6050 lb. It was an extraordinary combination of sophistication and barbarity, but the underlying objectives sometimes defied analysis. For example there was a two-speed reduction gear but only a single-stage single-speed supercharger, the idea being that the ratio of crankshaft to airscrew speeds could be varied in order to exploit the best characteristics of an airscrew in given circumstances. Something similar had been tried in the USA some twenty years earlier, when the test pilots used to enjoy themselves demonstrating their gearchanging skills in flight – until one day the gearbox was wrecked by a clumsy change-up during a power dive.

The intriguing valve gear of the big Lycoming provided another variable which had to be taken into account, together with airscrew efficiency factors and that invariable supercharger drive. Each of the nine cylinder banks had its own overhead camshaft, a most uncommon luxury in radial engines, and each shaft carried two separate sets of cams which were shifted to alter the valve timing, one giving maximum power during take-off, climb, or emergency conditions, the other promoting maximum economy in cruising. The idea is one that we have seen countless times, usually associated with engines for road vehicles, most recently in the attractively simple variable-timing camshaft developed by Associated Engineering Ltd. and most convincingly in the kinematic linkage devised by Torazza for Fiat and described in Chapter 3.

However, it would be a pity to give the impression that multi-bank engines were always failures. It may often have been the case (the ill-fated Rolls-Royce Vulture and Exe are two recent examples that spring to mind), but by no means always. It was therefore gratifying that one of the speakers at the IMechE discussion on unusual engines should have been that outstanding engineer Leo Kuzmicki, talking about the multi-bank tank engine of Chrysler. This was an engine of which 12 000 were produced, to be used in earnest during the war in North Africa, and it demonstrated an interesting and essentially practical approach to what was quite a difficult problem.

Between the wars the political climate created a situation where the United States found itself after Pearl Harbour in December 1941 without adequate facilities for making engines suitable for medium tanks of 25 to 30 tons weight. The only engine available in production and at all suitable was an air-cooled nine-cylinder radial aero-engine displacing a modest 16 litres and developing an inadequate 375 bhp. The big three car manufacturers therefore embarked on a programme to develop a suitable medium tank engine. Being the big three car manufacturers, they were not the types to allow engineering novelty to stand in the way of production convenience; but what emerged was an interesting combination of both.

The Chrysler Corporation started with an engine based on its successful in-line six-cylinder car engine already in production. Five of these were arranged around a central crankcase, with each crankshaft geared to a common ring gear, the total output being taken by a central shaft. The immediate advantage of this concept was the existence of most of the tooling. Many existing production parts could be capitalised including cylinder blocks, pistons, connecting rods, crankshafts, and cylinder heads. The major parts special to the design were the central cast-iron crankcase and the large end casting which housed all the gearing.

The configuration chosen resulted in a most compact layout, important because space in tanks is at a premium. The M4 tank was a rear-engined vehicle, each of its engines facing backwards from its normal position in a passenger car, and Chrysler's multi-bank layout ensured reasonable accessibility in this position.

The basic engine had a displacement of 4·12 litres and a stroke : bore ratio of 1·366 : 1. The compression ratio was lowered to 6·2 : 1 from the standard

6·8:1 simply by increasing the thickness of the cylinder head deck.

Production Dodge carburettors were used, one to each basic engine; but to allow each carburettor to operate in its normal position, a complex induction system with different tract lengths had to be evolved, with all but the uppermost bank having to use induction elbows to keep the carburettor in its normal vertical position.

Numerous problems emerged after the first prototype was run on the test bed: and the problems of carburettor synchronisation and equal fuel distribution were among the first. The very considerable difference in lengths between the individual induction pipes presented some airflow dynamics problems that Mr Kuzmicki described as 'fascinating, yet difficult'; but with the aid of inserted sheet metal vanes called centralisers, good fuel distribution was achieved and eventually all five carburettors were located in the same plane as the uppermost, which not only improved accessibility but also greatly simplified throttle and air cleaner arrangements.

Maintenance requirements also prompted the removal of each distributor from its normal position alongside the centre of the engine to a place at the end of the camshaft, which was driven by gears instead of the usual chain. Early experimental engines retained separate water pumps, each driven by a belt from the crankshaft pulley – but the absence of a flywheel on each individual engine resulted in extremely short belt life. Maximum reliability being essential for the envisaged duty, the engine was redesigned to incorporate a single central water pump driven by an accessory shaft in the central crankcase: this pump provided cooling water at the end of each cylinder block, with an internal distributor tube ensuring proper circulation within the engine. In fact cooling and the effective sealing of the system, despite the problems of Siamesed cylinders, required a good deal of development work because of the heavy-duty operating cycle typical of tank engines. The same needs prompted the use of a chromium-plated piston ring and of dry sump lubrication with an oil cooler in the circuit. Mechanically there was another unusual refinement in that the output gear train of the engine had to have its five pinions and ring gear mounted in what was virtually a self-contained gearcase assembly in order to ensure proper gear alignment. To maintain the necessary axial tolerances, splined sleeves were used to transmit the torque from the flange of each crankshaft to the corresponding pinion in the gear train assembly – and with such complicated tolerance combinations, noise was a big problem for a long time.

As finally developed, the multi-bank engine developed about 470 bhp at 2700 rev/min. Maximum torque occurred at the rather low speed of 1200 rev/min and was 1070 lbf ft. Both these performance factors were a good deal

Wartime ingenuity: the 1943 Chrysler Multi-bank tank engine employed a high proportion of parts common to their private car production. The five banks of cylinders can be clearly seen in this photograph.

Courtesy: Chrysler Corporation.

1943 Chrysler Multi-bank tank engine. The end gearing arrangement is illustrated in this view.

Courtesy: Chrysler Corporation.

better than the aforesaid aero engine could offer, though no doubt that in turn was a good deal lighter than the 5244 lb accounted for by the multi-bank Chrysler complete with radiator, cooling fan, and clutch.

The tank seems to have taken over from the aircraft as a home for odd-shaped engines. In another chapter the 90° V10 air-cooled Mitsubishi affords another example; and another air-cooled two-stroke, intended like the multi-bank Chrysler to propel US army tanks, was the Orion. The requirement that prompted this design was for an air-cooled compression-ignition engine of compact dimensions and modest weight with an output of 600 bhp. A space of just 1·7 m^3 was available for the engine, implying a specific volume of less than 0·003 m^3/hp. The result was an opposed-piston two-stroke engine with four crankshafts moving twelve pistons in six cylinders in two layers – and to make the whole thing more improbable still, the crankshafts were only there to keep the whole thing together, power being taken from the shaft of an exhaust turbine for which the piston engine acted merely as a gas producer. In these circumstances the Orion engine is more appropriate to review in Chapter 5; but it is worth mentioning here because of its place in a class of engines that is surprisingly large and that is nevertheless considered perpetually unusual – that is to say the opposed-piston two-strokes.

Of these, the three-cylinder six-piston Rootes compression-ignition engine designed for truck and 'bus duty is one of the best known and most widely used. Its most unusual feature is the use of rocker arms connecting the piston connecting rods to larger and more conventional connecting rods swinging about the pins of a conventional crankshaft located below the cylinder block. The arrangement has the virtue of reducing the overall width of the engine, and of making such adjustments to primary and secondary connecting-rod angularities as may be desirable by appropriate alteration to the rocker arm ratio and the crank throw (which need not of course be the same as the piston stroke). However, the layout lends itself to experimentation in so many directions that this engine has become the most favoured workhorse of research teams at the British universities, including the compound differential diesel of Professor Wallace discussed in Chapter 1. The Timoney engine is another such variant, developed in the mechanical engineering department of University College Dublin by their head of thermodynamics Dr S G Timoney. His object was to produce a diesel engine of high specific output, compact dimensions, and exemplary durability; and the crux of his work lies in the simple exploitation of the usual rocker arms to give rapid and precise control of a variable compression ratio.

There are two interdependent principles involved in the Timoney engine. One is the pursuit of high specific output through the combination of a two-

Courtesy: Engineering.

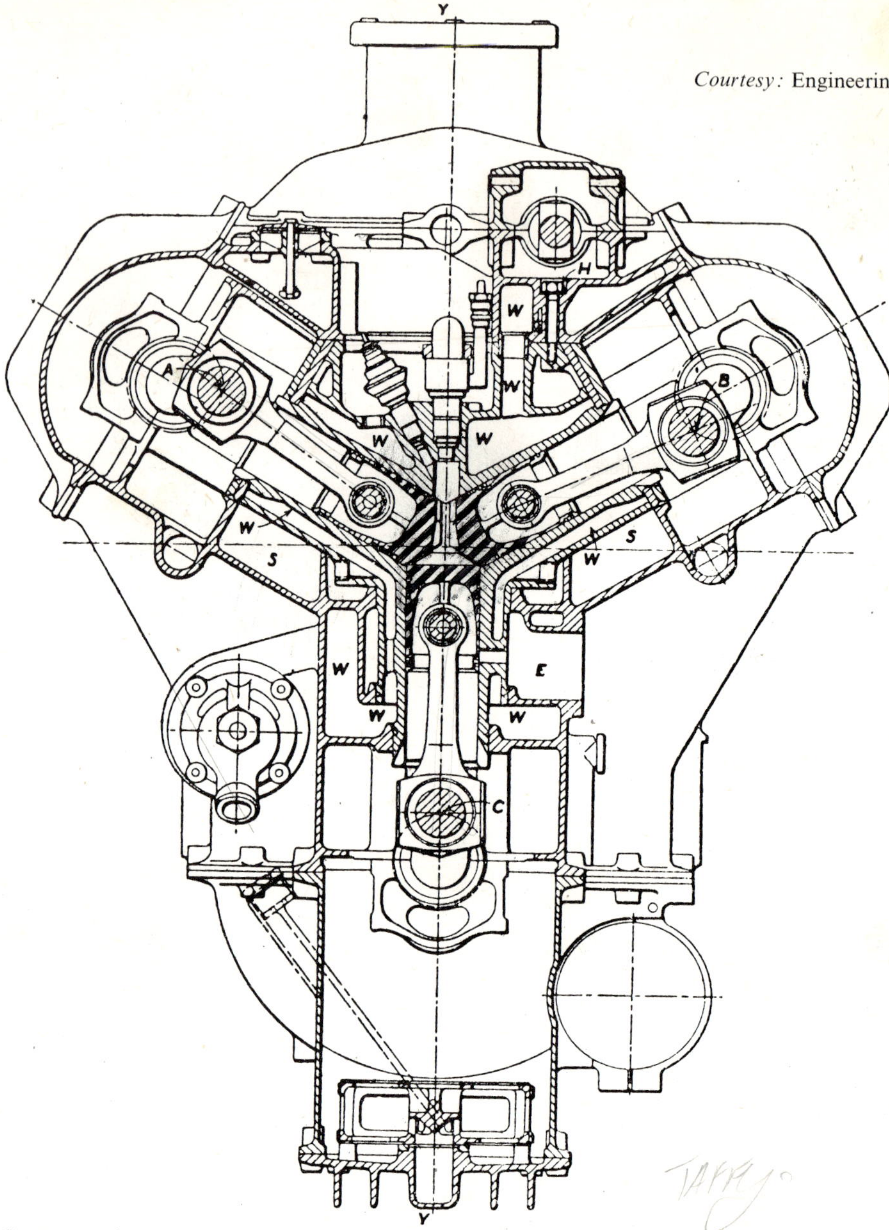

All for one and one for all: The 1920s Michell three-cylinder two-stroke compression-ignition engine took the opposed-piston principle one step further in that the three cylinders forming each star-shaped unit (multi-cylinder versions were designated two-star, three-star, etc) shared a central common combustion chamber, each piston operating through normal connecting rods of equal size to three separate crankshafts. The crankpins were coupled to a duralumin triangle (which also enclosed a novel scavenging pump) to ensure that all three shafts rotated at equal speed.

The passage of the scavenging air was assured a direct path through the cylinders by virtue of the arrangement of the valve porting, the two upper cylinders of the Y formation controlling the inlet ports and the lower one controlling the exhaust. *(continued opposite)*

The very fact that the opposed-piston three-cylinder diesel two-stroke was capable of very high output, under boost conditions, was itself the cause of further problems which, for a time, threatened its further development particularly in relation to the Wallace compound cycle.

Unfortunately, the willingness of the engine to return performances possibly beyond the intentions of its original designers led to serious failure of the bearings under sustained boost conditions. This tendency was overcome by the introduction of the Timoney system of variable compression ratios which, by virtue of a system of rocking beams, control levers and damper springs, enabled the engine to become self-governing, so far as compression ratio (and hence peak pressure and consequent bearing loads) was concerned.

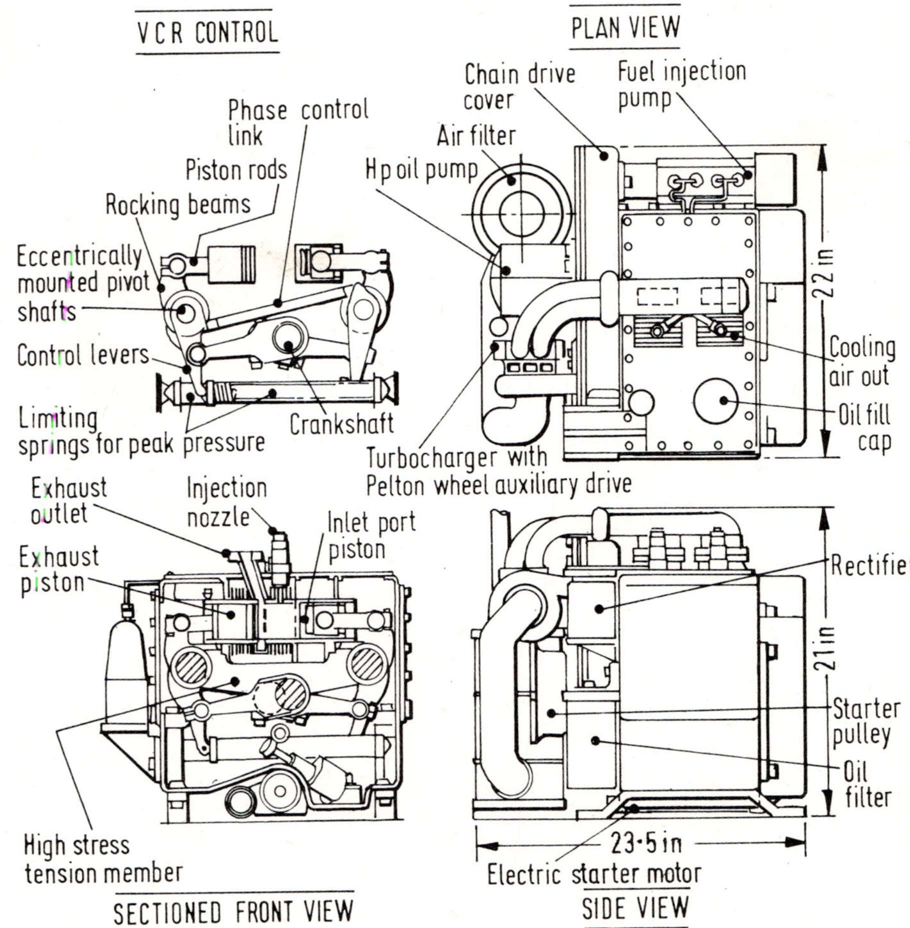

stroke cycle and exhaust gas turbocharging; the other is the automatic variation of compression ratio to ensure long service life by controlling loads on the pistons. Thus part-load operation may have its efficiency increased by raising the compression ratio, while the high combustion pressures attendant upon full-load operation and a high boost pressure are suitably modulated by lowering the compression ratio. The mechanical means of achieving this are very simple, relying on eccentric bushes for the pivots of the rocker arms.

Clearly the change in compression ratio should be in response to changes in engine stress, which is equivalent to changes in peak cycle pressure. This makes it possible for the control mechanism to be made automatic by attractively simple means. The eccentrics are rotated by levers which, if left unrestrained, would tend to turn under the load supplied to the pistons by combustion pressure. A spring link between them provides the necessary restraint and responds directly to the cylinder pressure, adjustment of the spring rate controlling the amount of change of the compression ratio. A phase control link ensures that both pistons move by the same controlled amount, lest all the change occur at one piston only; otherwise there would be no positive assurance that the intended change of compression ratio, and consequently of the timing of the piston-controlled ports, could be maintained.

How fast is the response? Faster than could be required of an engine in service: artificially rapid loading and unloading of the engine on a test bench proved that the compression ratio could be changed from a minimum of 9·5 : 1 to maximum of 16·5 : 1 in six cycles of the crankshaft, and commensurately reduced in eight. As a consequence of this promptitude it proved impossible to produce audible diesel knock in the Timoney engine. The interesting and rather unexpected corollary of this is that the load on the eccentric shaft control levers, and hence on the spring, must be a function of the peak cylinder pressures and not of the mean pressure as might have been expected. This has been borne out by investigation confirming that the system performs precisely the intended function.

The virtue of the variable-compression-ratio system is that it enables an engine to have designed into it suitable characteristics for a wide range of applications in marine, vehicle, and stationary duties. Because of the control of peak cycle pressure it is practical to run at bmep values double those of conventional engines of similar size and weight. Heavy loads are intercepted before they can be transmitted to the running gear bearings, so their lubricating oil film wedges remain intact; and a similar immunity from transient overloads applies to all other engine components, ensuring the long life that was one of the aims of the exercise. Of course there are other ways of achieving controlled variation of compression ratio, but it is difficult to think of any as elegant as this system, made possible by the unusual format of the Rootes two-stroke.

If the opposed-piston two-stroke is automatically to be deemed unusual, the Napier Deltic must be the most unusual of all. Despite its unconventional layout – perhaps because of its elegance – it has been by no means unsuccessful, being employed for rail traction, electricity generating, pumping, and for marine use in small high-speed naval craft, motor yachts and postwar minesweepers for which the engine's high power : weight ratio and freedom from the need for skilled maintenance work make it eminently suitable.

The Deltic is an engine (or rather a range of engines) of eighteen-cylinder construction, the cylinders being arranged in three banks of six to form an equilateral triangle. A six-throw crankshaft is located at each corner of the triangle, each crank throw carrying a forked and a blade rod. The forked rod carries the exhaust piston, the blade rod carries the inlet piston; and because of the geometry of the layout, the two pistons work at a phase angle difference of 20°. This exhaust piston lead is clearly advantageous in promoting efficient scavenging of the cylinders which, as already explained, are of the opposed-piston compression-ignition two-stroke type, liquid cooled and set in a structure that is substantially of light alloy. All the crankshafts are of course coupled through phasing gears, the lowest running in the reverse direction to the other two.

The first generation of Deltic engines had mechanically driven scavenge blowers and produced 2500 hp at 2000 rev/min. By the standards of 1950 this was most impressive, as was the maximum continuous power rating of 1875 at 1700 crankshaft rev/min allied to a peak bmep of 92 lbf/in^2 and a specific fuel consumption of 0·422 lb/hp h. By 1958 the second generation acquired a turbine-assisted scavenge blower, with maximum and rated powers, bmep and sfc, improved respectively by 24, 28, 18 and 5·5 per cent respectively. After a similar interval a third generation of higher horsepower came along, basically similar in design to the earlier units but with a redesigned turbocharger of increased efficiency, and with charge air coolers interposed between the blower and the cylinder blocks enabling a greater mass of air to be trapped at a lower temperature with the charge pressure remaining unchanged. By these means the power of the engine for fast patrol-boat duty was increased to 3700 bhp at 2100 crankshaft rev/min, making the peak bmep no less than 129·6 lbf/in^2 for a penalty in sfc of only 0·75 per cent. Perhaps inevitably the weight suffered, going up from 11 000 lb in 1950 to 15 750 lb in 1966 so that the weight : power ratio (including the reverse/reduction gearbox) only improved from 4·4 to 4·25 lb/hp.

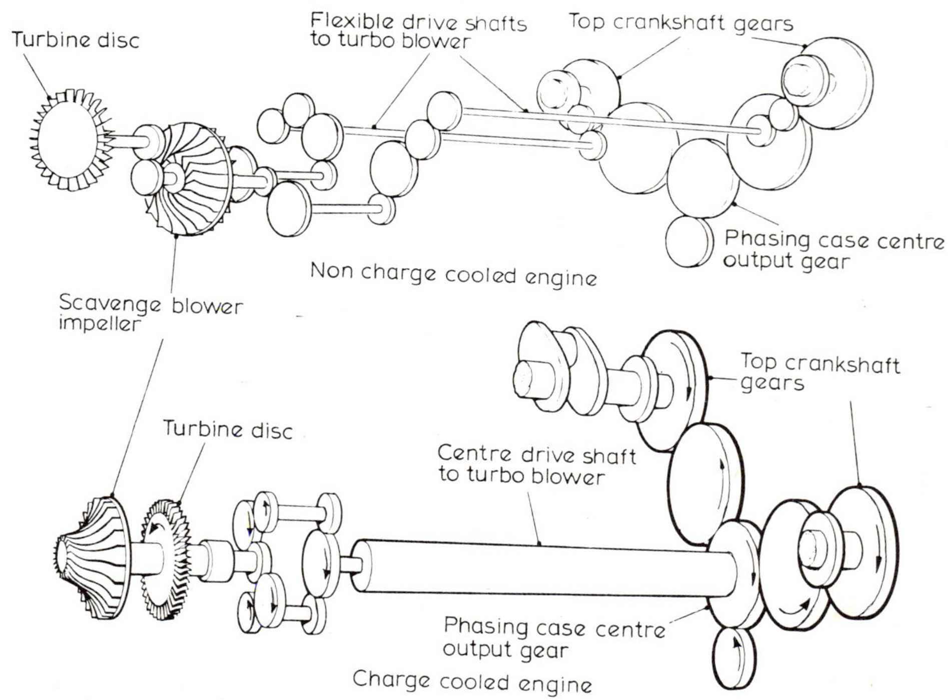

Non charge cooled engine

Charge cooled engine

Taking its name from the Greek letter Δ or Delta, which it echoes in shape, the Napier Deltic has three banks of cylinders set on the flanks of an equilateral triangle with crankshafts at its apices. It thus anticipated the geometry of the Lear Delta 'vapor power' (sic) steam engine developed in 1969.

End gearing was employed to transmit the power from the three crankshafts to a centre output shaft, and since the triangular island formed between the cylinders made a convenient home for it, and since the centre gear in the phasing casing was a nodal point, a centre drive shaft was taken from it to a forward turbo blower. The space between the cylinders also accommodated a speed-increasing gearbox between the centre shaft and the blower, providing a step-up ratio of 5.16/1. Compact though the arrangement was, Napiers no doubt found (as did Lear Motors Corpn) that it provided inherent difficulties in accessibility and maintenance.

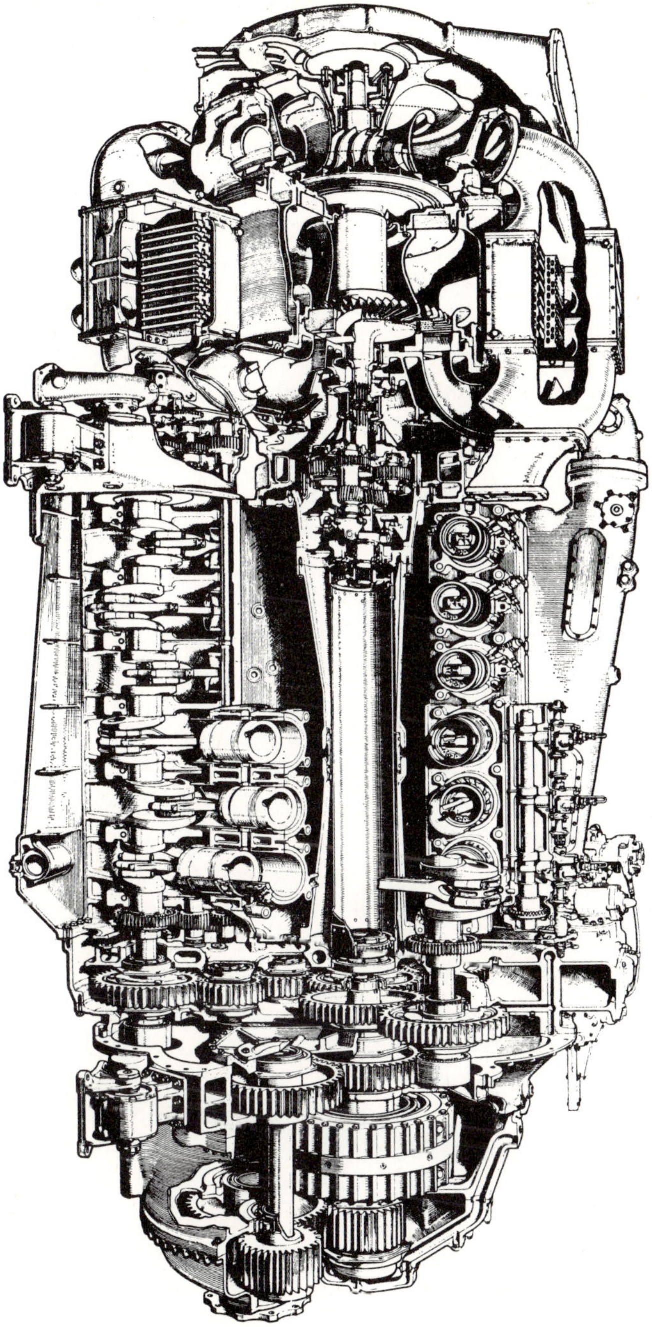

One of the most attractive details of the third generation engine was the means for positively driving the turbo-blower from the crankshaft. In earlier versions of the Deltic, this drive had involved torsionally flexible shafts, but when the engine was idling in neutral gear at low temperatures an adverse torsional vibration occurred in the drive which led to premature failure of the shafts because of torsional fatigue. An extensive theoretical investigation of the drive showed that the centre gear in the phasing trains that linked the crankshafts was a nodal point and under such conditions an ideal position from which to take the drive from the turbo blower, simplifying the transmission enormously. A very simple speed-increasing gearbox provided the necessary step-up ratio from the large-diameter central drive shaft and incorporated two layshafts carrying helical gears.

These layshafts were free to move axially, each bearing against a balance arm through barrel bearings. Because of the helical cutting of the gear teeth, the axial force in each layshaft was a function of the torque transmitted; and the centrally pivoted balance arm ensured identical torque transmission to each layshaft, eliminating the need for phasing of the transmission gears by vernier adjustment or other means. It was a piece of typically elegant Napier design, recalling an essentially similar arrangement they employed in their prodigious Sabre aero engine to compensate automatically for any phasing differences between the two crankshafts while taking their drive to the central output shaft. It is sad to think that if BRM had taken the opportunities given them to study the Sabre, they would have been saved nearly all the serious difficulties they encountered in the development of their H16 racing car engine in 1966–7 and might well have emerged from those seasons with the most powerful Grand Prix engine in regular and reliable use.

Alas, it is too easy to be carried away by the consideration of what might have been, large though this element loomed in the original conception of the Institution's discussion. In retrospect, perhaps the most significant feature of the programme of that meeting was that, although all the engines discussed must be presumed unusual (whatever one might think of the essential ordinariness of the Hassan and Mundy designs), not one of them was so daringly unconventional as to dispense with conventional pistons. This was perhaps a proper reflection of history, which teaches that so long as an engine has ordinary understandable pistons and cylinders it can be as unusual as it likes and still have a fair chance of success, whereas to dispense with these essential elements is evidently to invite commercial failure no matter how great the technical promise.

Look at all the unholy rabble of rotary engines that have been spawned by idealistic inventors over the years. We all know about the Wankel, which is

Upstairs, downstairs: the upper and lower crankshafts of this 1935 Junkers Jumo aero engine made for a narrow vertical profile with small frontal area. An opposed-piston diesel two-stroke employing Oechelhauser principles, the Jumo was put into limited production but achieved little success. The end-gearing linking the upper and lower crankshafts to the propeller shaft can be seen to the left of the photograph.

Courtesy: Science Museum, London.

Although the James employs opposed pistons operating from a single crank with a common combustion chamber, it can in no way be said to resemble even remotely the Gobron-Brillié design. Current developments include a stratification chamber on Ricardo principles in a twin crank configuration.

Courtesy: Richard James Cylindrical Motors Ltd.

Pistontravel on James opposed piston engine

PISTONS ON T.D.C.

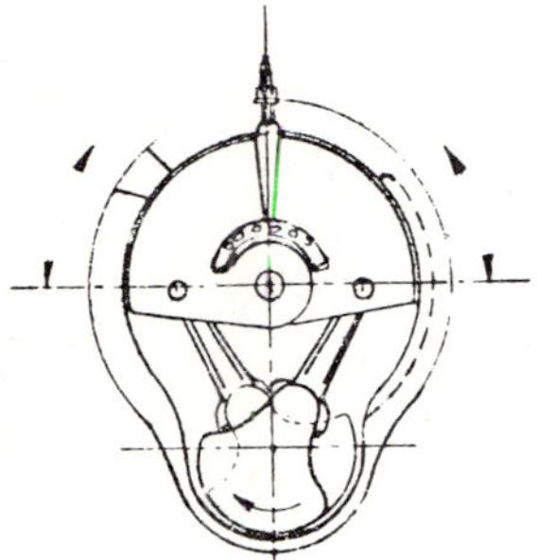

FIGURE: 1

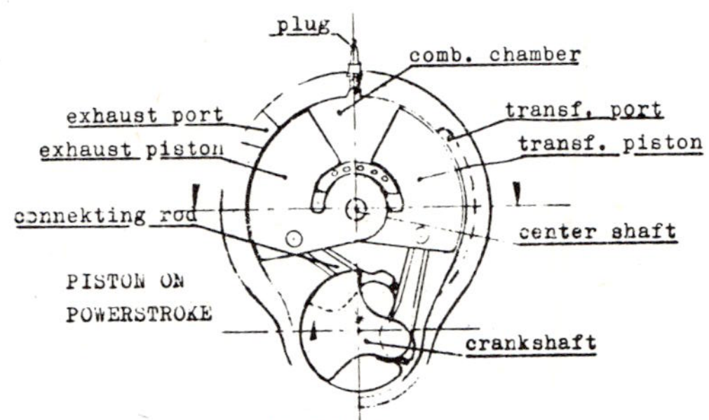

PISTON ON POWERSTROKE

FIGURE: 2

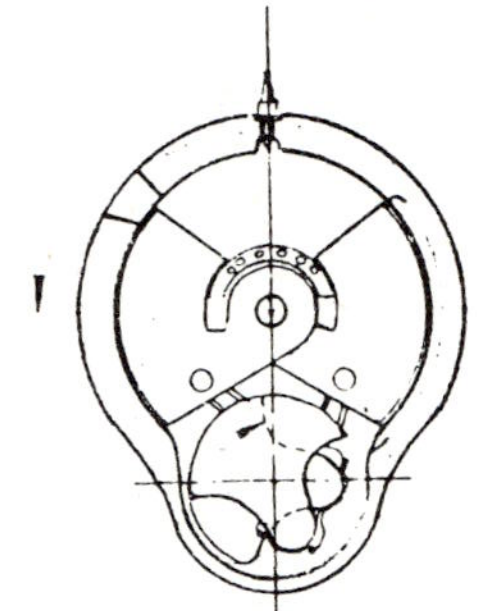

FIGURE: 3

EXHAUST PISTON OPENS FIRST ON POWERSTROKE

PISTONS ON B.D.C.

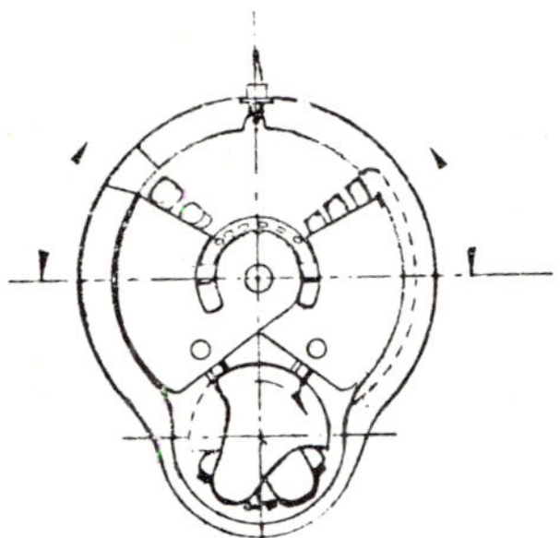

FIGURE: 4

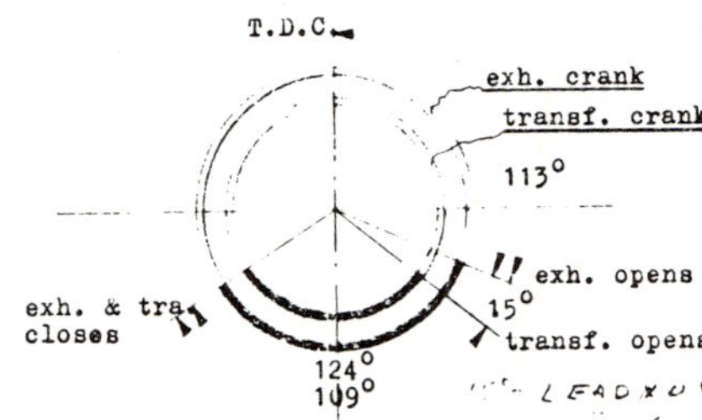

port timing diagram

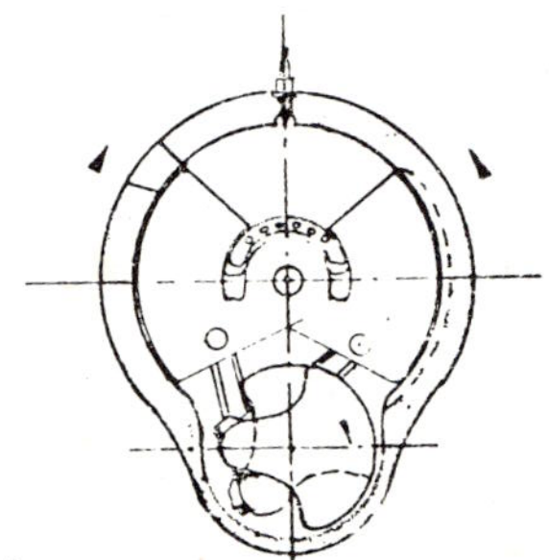

FIGURE: 5

EXH.- AND TRANSFER PISTON CLOSE AT SAME TIME ON COMPR. STROKE

Spark plug and fuel injection in
pre combustion chamber

in large scale production even if we do not know whether it will come to predominate or come to grief in ten years' time. But what of all the other trochoidal rotor configurations, the Isuzu with a three-lobe housing, the Variant with four lobes, the Renault-Rambler with five, and so on? What of the multitude of rotary engines based on the old eccentric-vane pump concept, culminating in the 400 hp 188 lb Mallory with its continuous combustion and dipsomaniac thirst for oil? They are lost causes, every one, always improbable and seldom plausible – nine days wonders, including the very professionally publicised Sarich which captured the uncritical imagination of far too many newspaper correspondents in 1973.

Then there are the assorted engines with 'clap-hands' or 'cat-and-mouse' pistons. In some of these, such as the Tschudi and the Bradshaw Omega, the pistons are very nearly conventional but move in arcuate paths through matching toroidal cylinders. Because of the circular cross section of the toroidal track, compression can be maintained by conventional piston rings, while the pistons chase each other around at varying distances to achieve the compressions and expansions with which we are familiar. In the Tschudi engine there were four such pistons, fixed in pairs to concentric shafts whose phasing was controlled by a set of rollers and cams; whether it worked is not known, but the essentially similar Bradshaw engine certainly did not. On the other hand the Kauertz engine, employing wedge-shaped pistons but otherwise similar in principle, was reported to have produced 213 bhp on the bench; while another, invented by Melvin Rolfsmeyer and known as the Virmel, achieved 337 bhp from a displacement of 2·43 litres and a weight of only 130 lb.

Perhaps the most unusual of all engines with clap-hands pistons is the one that retains conventional connecting rods and crankshaft. This is the James opposed-piston engine, which stems from the work of Richard James of Canada. As far back as 1929 he sought to design an opposed-piston engine with only one crankshaft and with conrods going direct from it to the pistons. After much work he gave it up as impossible; but in 1963 he revived the idea, recognising the many advantages of the opposed-piston concept, and this time succeeded in producing an exceptionally compact design.

The James engine has two opposed pistons substantially wedge-shaped and pivoting about the centre of a drum-type cylinder. An extension of this cylinder chamber houses the crankshaft, with connecting rods linking it to the pistons which can be made to travel at different speeds relative to each other during different phases of the stroke. Thus the exhaust port can be uncovered ahead of the intake port and close either at the same time as or ahead of the intake port. This can prevent loss of mixture or, in the case of a compression-ignition engine, loss of scavenging air; and if the exhaust port be the first to close, it allows supercharging. As one piston controls the exhaust port and the other the transfer ports, a uniflow scavenging of gases is achieved which in turn allows greater port areas with a minimum of port height, resulting in a greater effective stroke. Moreover, the uniflow scavenging system used in this engine involves long gas paths while the ports are not at all close to each other: the result is clean scavenging at low rotational speeds, making the most of one of the characteristics of a piston engine which is high torque at low crankshaft rate.

Other features of the James engine include low friction losses, for the pistons oscillate around a centre shaft without actually bearing on the cylinder walls, leaving the seals with purely sealing duties. However, the most obvious advantage of the type is its compactness: compared with other industrial engines (this being the most promising application) the James engine is smaller and lighter for a given performance, its bulk and mass being comparable with those of a Wankel engine but its potential for combustive efficiency and fuel economy considerably better.

The James engine has earned some favour and support from the Canadian Federal Government, enabling most development work to be virtually completed, although there remains the possibility of the design being adapted for stratified charge operation. Whether it will go into production remains to be seen, but it certainly ranks as one of the most unusual designs to be numbered among the engines that might well be.

5. Engines with Unusual Translation of Power

Can one imagine the internal combustion engine having a place in an ideal world? If that be possible, it is certain that if it had connecting rods they would be of infinite length. As it happens, Nature appears to be somewhat at variance with the aims and objects of the engine designer, who is accordingly often tempted to dispense with connecting rods and all the expensive and problematic paraphernalia that are associated with them. Alternatively, if he seeks to retain the piston rod, he may be at some pains to eliminate the crankshaft – or at least, as in the case of the scotch yoke favoured by such eminent designers as the late Stuart Tresilian and the still active Philip Vincent, to get rid of the disturbing angular accelerations of the connecting rod. Unfortunately the engineer's quest for some alternative means of translating the motion of the piston to that of a rotary output shaft has generally led to his undoing.

The axial engine provides ample evidence of this. It is an arrangement that has charmed designers for many years because it is blessedly compact, with the cylinders of the engine arranged in the same way as arc the chambers of a revolver. With an output shaft at the centre of this cylinder block, all that is needed to link it to thc pistons is a swash plate or wobble plate. Could anything be simpler or more obvious?

Could anything be more difficult? Not if the record of umpteen axial engines built and abandoned as failures is any guide. Yet some of them worked, and in some respects worked well; and none of the difficulties encountered with them should be beyond the capacity of current technology to solve.

As long ago as 1875 a steam engine was designed on these lines, a swash-plate affair patented by one D K West. By 1912 there was an axial petrol engine propelling an American car: this was the Macomber, a five-cylinder engine that was a true rotary in that the reaction member was held stationary while the rest of the engine – about 95 per cent of its total mass – revolved. Smooth though it was reputed to be, the Macomber did not survive; and neither did the many other axial engines propounded in the first thirty years of the century – the two-stroke Lind, the Smallbone, the Almen, the Nagelmann, the Gerardine (which was an hydraulic motor, paving the way for the successful hydrostatic axial motors of today, none of which works satisfactorily at other than low rates of rotation) the Redrop, the Helm, and so on.

This did not deter Major C G Nevatt, one time President of the Institution of Automobile Engineers, from the temptation to produce a wobble-plate engine in the factory of the Bristol Tramways and Wagon Company where he was general manager of the motor works in the early 1930s. At the Institution's 1971 discussion meeting on unusual engines, George White described how this design was prepared. It was a nine-cylinder of seven litres displacement, the principal feature being a Z crankshaft.

This type of crank, suggested as long ago as 1905 by J G Maxwell, overcomes some of the snags associated with the swashplate arrangement, notably the inherent lubrication difficulties and the inevitably large size and mass of the plate itself. Instead, a shaft with an inclined crank journal is used; and upon that journal is borne a sloping but symmetrical member to the extremities of which the piston rods are connected by ball joints or something equally universal in motion.

The movement of the spider member that links the piston rods to the crankshaft is less ideal than might be hoped. Any point on it moves not in a straight line but in a figure-of-eight path, two complete circuits of the figure being completed for each revolution of the crankshaft. The path of one of the arms of the wobble plate when viewed in end elevation is a perfect circle. Accordingly there is no hope of providing straight-line motion for the piston rods and eliminating a joint or two; and what proved even more troublesome, according to White, was the need to support or anchor the wobble plate to prevent it from rotating, so that it might transmit the torque reaction to the crankshaft instead of simply tying knots in the piston rods. Were the piston rod motion in a straight line, the problem would simply disappear with the adoption of guides for the rods; instead Bristol had to make something rather like a frying pan, the handle pointing downwards and mounted in a bush fixed to the crankcase.

After three weeks' bench testing which seemed to go very nicely, the engine was stripped for examination, and it was found that the cast-steel wobble plate was about to collapse. A new one was made from roughly forged billets and stabilised by two bevel gears, one mounted on the wobble plate and the other on the crankcase. At the same time a dry sump system of lubrication was adopted and this caused trouble. When the oiling system was redesigned, the bevel gears were scrapped and the frying pan readopted,

Shown here on dynamometer test, the compactness of the Bristol axial engine allied to its increased potential power output (or a saving of 16 per cent on fuel for comparative output) when set against conventional units, gave its sponsors high hopes for its success. They were not fulfilled.

Courtesy. Bristol Commercial Vehicles Ltd.

but the handle was reciprocated at twice engine speed at right angles to the Z crankshaft . . . and this was a failure too.

So was the rotary valve that Bristol tried. The original engine had poppet valves, but the layout of an axial engine constitutes an open invitation to the adoption of a simple rotating disc carrying passages that can link a single cylinder port to the appropriate inlet and exhaust tracts in due order. Bristol took to the idea quite readily and produced an engine with admirable breathing arrangements. These were justified by the performance figures: the engine developed 150 bhp at 3000 rev/min, compared with about 115 for the 7-litre engine of conventional design that it was intended to supplant. The torque was greater (320 lbf ft compared with 294) and the overall fuel consumption about 15 per cent lower, the sfc curve remaining consistently below that of the conventional engine at speeds up to about 2200 rev/min.

Unfortunately the rotary valve was not a success, owing to the impossibility of maintaining a seal on its operating base without flooding it with oil. This is a story we seem to have heard elsewhere – but in any case it is quite likely that the improved performance of the wobble-plate engine was due to a reduction in frictional losses. The engine needed only about one-third of the power to motor it around that was demanded by the conventional equivalent engine.

Perhaps the most successful example of the axial engine is a true swashplate type, the Michel, which uses the eponymous thrust pad of marine practice to reduce the frictional losses caused by the thrust on the inclined plate. It can be made either for a two- or four-stroke engine, Michel doing a two-stroke diesel that gave a bmep of 95 lbf/in^2 at 900 rev/min. Balance is no great problem, all that is necessary being to embody at regular angular intervals at least three pairs of opposed cylinders.

As well as the swashplate and wobble-plate engines, there is another class of axial types using a revolving cam plate to reciprocate the pistons, which can conveniently be double-ended to work in opposing cylinders. The beauty of this arrangement is that the piston motion is in a straight line, and if the cam plate is generated in sine-wave form the motion is truly harmonic. When the cam profile comprises two sine waves, one revolution of the cam plate and output shaft corresponds to four strokes of each piston, which may be a complete four-stroke cycle or two two-stroke cycles.

Two engines of this type are worth noting, both of them American. The Alfaro was made by the Aircraft Development Corporation of Boston, and had four cylinders giving a displacement of 2·51 litres. Its maximum output of 115 bhp and corresponding bmep of 136 lbf/in^2 at 2000 rev/min were not remarkable, but the virtues of the axial engine shape for some applications

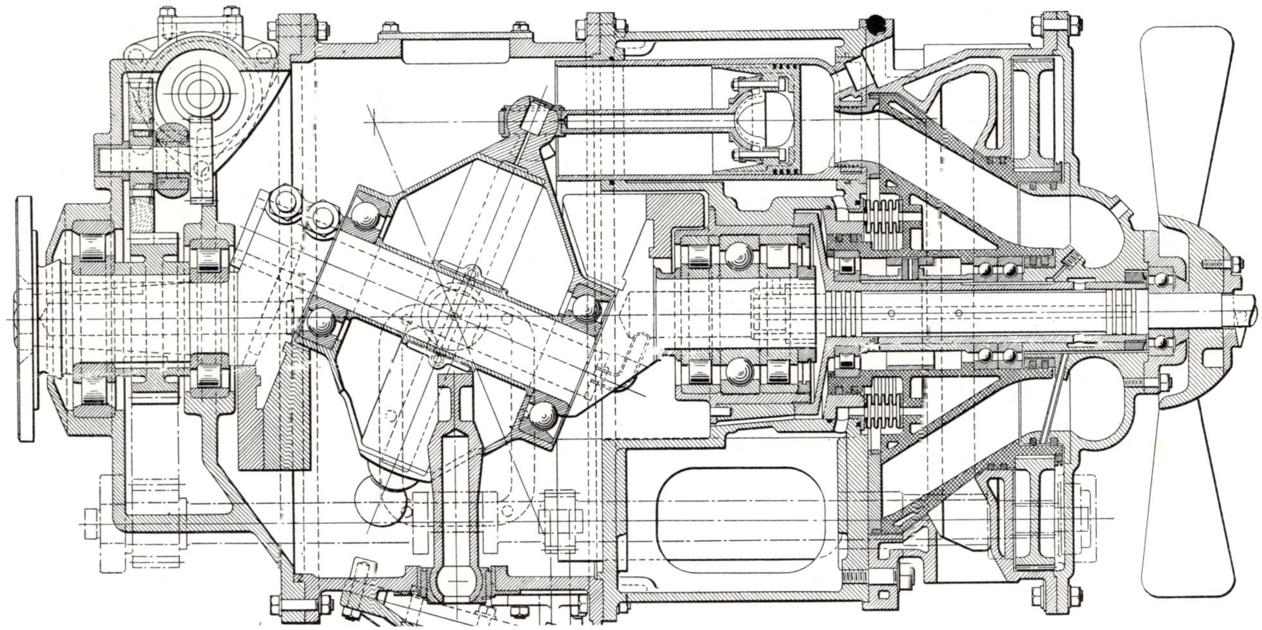

The Bristol axial (or 'wobble-plate') engine of the mid-1930s was originally conceived as a commercial vehicle prime mover by the old Bristol Tramways and Carriage Company, with bus and coach operation particularly in mind. This section drawing illustrates the arrangement of the 'wobble-plate' on its 'crankshaft' quite clearly.

Courtesy: Bristol Commercial Vehicles Ltd and Dr E S Moult, CBE.

(notably light aircraft) were given eloquent testimony by the figure of 86 bhp for each square foot of frontal area. Similar virtues were demonstrated by the Herrmann, a twelve-cylinder cam plate engine furnished with the poppet valves normal for a four-stroke. This gave 144 bhp at 1600 rev/min and 120 lbf/in^2 at the peak of the bmep curve, combined with a respectable fuel consumption of just 0·5 lb/hp h. When supercharged, the maximum output rose to 245 bhp on 100 octane fuel, and during a twenty-five hour endurance test the engine sustained between 208 and 213 bhp. It weighed only 249 lb, was 30 inches long and 12·7 inches in diameter – but as well as the respectable ratios of power to weight and to frontal area, we meet again the claim for low frictional losses, the manufacturers declaring that these amounted to only three-quarters those of conventional engines of similar capacity.

There is no doubt that the axial engine is very attractive. It is compact, especially low in frontal area, light in relation to its output, and very simple in general arrangement (including auxiliary drives). Moreover its balance can be made perfect by appropriate attention to the number and disposition of the cylinders, and in the case of the cam-plate variety it is completely free from secondary vibrations. On the other hand, lubricating problems are rife – not only due to the difficulty of contriving suitable bearings to take the piston thrust, but also because of the centrifugal action of the swashplate or cam-plate on the oil. Severe wear at this point has been remarked in a number of instances, and there have also been complaints of a general want of accessibility to the principal working parts for normal maintenance. The fact that maintenance such as was considered normal in the days when these engines were experimented with would be considered excessively pragmatic today may offset this last objection.

However, engineers have found plenty of alternative ways of doing away with the crankshaft, the connecting rods, or both. Not the least popular has been the idea of using the piston engine proper as a gas producer whereby to feed a turbine whose shaft constitutes the power take-off point. The Orion tank engine mentioned in chapter 4 was one such; connecting rods and crankshafts it had in plenty – twelve of the one and four of the other – but they had nothing to do with power output. This was drawn from an exhaust turbine fed from the piston-controlled ports of the six opposed-piston two-stroke cylinders.

The most remarkable feature of the original Orion design was that not only the combustion gases but also the vitiated cooling air were to be fed to the power turbine. A crankshaft-driven centrifugal impeller provided air for scavenging and for cooling, and the idea was that the latter quantity would (after passing over the fins on the outside of the cylinders and being there

1947 Wooler Swashplate engine. Better known for his unorthodox motorcycle designs, John Wooler made this aero engine similar in many respects to the Bristol axial (or wobble-plate) engine and the term 'swash-plate' is a misnomer. It differs in that *two* wobble-plates are employed, activated by an opposed-piston layout. The 12 pistons are arranged in six cylinders with their longitudinal axes parallel to the central propeller shaft and with six centrally-located combustion chambers.

Courtesy: Science Museum, London.

heated) be ducted to the turbine through the same channels as the combustion gases. It was very difficult of course to provide the appropriate division of flow inside and outside the cylinders, though the dense finning necessary to achieve the intense cooling demanded by the high rating of the engine produced enough resistance to air flowing over the outside of the cylinders to encourage sufficient to go inside. Nevertheless it was eventually necessary to abandon the original idea of using the same impellor for scavenging and for cooling air, since it was impossible to obtain the same flow resistance in the ports and cylinders as in the finning, and also because of the need for thorough filtration of the induction air. The amount of air required for cooling was three and half times as much as was needed for combustion; any cleaner able to cope with such a flow efficiently would be far too bulky to house in a military tank, where space requirements were so straitened that only $1{\cdot}7\ m^3$ were allowed for the whole engine.

The Orion encountered numerous other snags as well. The spray of oil on the underside of the piston crowns cooled them so well that more than 4 per cent of the total calorific value of the fuel was accounted for in this way – and when it is remembered that the heat removed by air cooling from the cylinder finning was not lost but was used in the turbine it may be thought a trifle extravagant that the major heat loss should be through the oil cooling of the pistons. Uncooled pistons were therefore tried but failed in less than five hours at full load, and there were associated difficulties with piston ring and gudgeon pin lubrication. At any rate, the Orion never got beyond the prototype stage.

There was another kind of opposed-piston two-stroke gasifier engine that went into service in quite a big way, and that epitomises better than any other the art of making an engine whose pistons have no connection with the output shaft. This was the free piston engine, discussed during the Institution's 1971 meeting by Professor Wallace whose work on the compound differential opposed-piston diesel has been noted in chapter 1.

As Professor Wallace told his audience, the free piston engine represents at any rate in theory a most elegant solution to the problem of obtaining extremely high output from a diesel engine; and by using the exhaust turbine rather than the non-existent engine crankshaft to provide the drive, it overcomes at the same time many of the disadvantages inherent in the use of piston engines as prime movers, notably the intermittent nature of the torque and the severe limitations on torque back-up for traction duties.

The idea was put forward by Pescara in the late 1920s, and was developed almost entirely in France, though a number of French designs were later built in Britain. Development was virtually continuous from 1922, except for

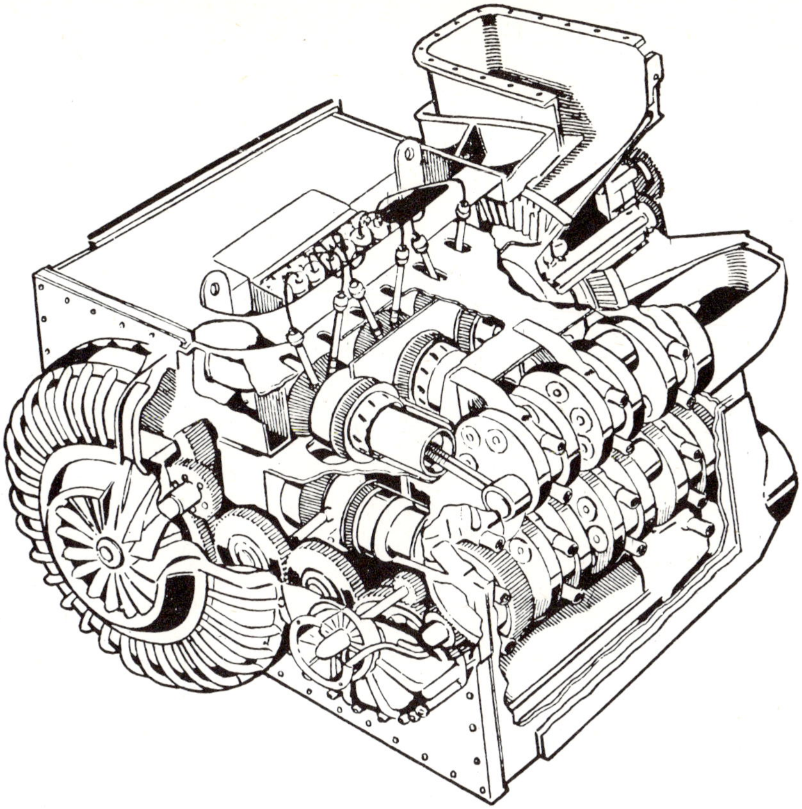

Slain not by Diana, but by General Motors themselves, the Orion two-stroke engine was unfortunately, like its mythological predecessor, attended by stormy weather. Basically two horizontally-opposed air-cooled sixes superimposed one upon the other, the 12-piston compression-ignition Orion drove four geared crankshafts sharing a common crankcase.

Its chief interest, however, lay not in its geometry but in its attempt to utilise the heat energy of the cooling air by passing it through a turbine. Air emerging from the supercharger was split into two streams, one of which passed through the cylinders, and the other over the cooling fins. It was found that the engine could be cooled by air with a temperature of 115 degrees centigrade at a pressure of 2 kg cm^2. Difficulty was experienced in cooling the pistons, however, oil spray cooling of the underside of the piston crown resulted in heat loss which could not be recovered. Finally the idea of using the same supercharger for scavenging and cooling had to be abandoned, since it was impossible to obtain an equal flow resistance in the ports and cylinder as in the finning, and also it was burdensome but necessary thoroughly to filter induction air.

Courtesy: General Motors Corporation.

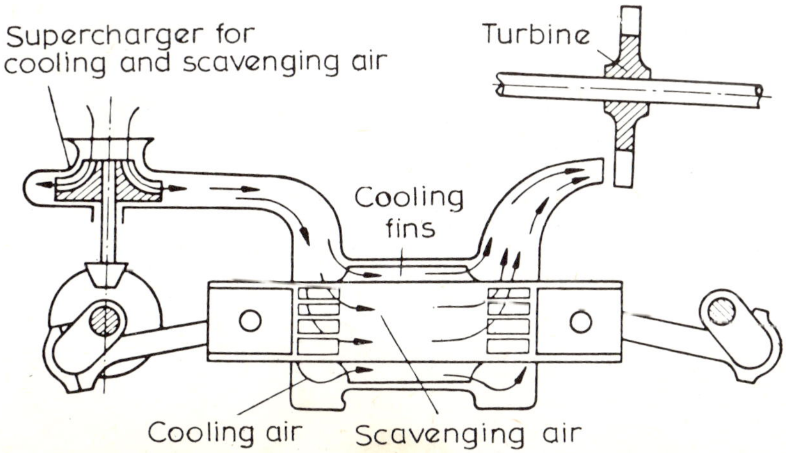

the war years; but little came of it until the perfection of the gas turbine developed a need for the hot moderate-pressure gas produced by the pressure-charged compression-ignition engine.

Two organisations stand out in the development and manufacture of complete free piston engines: SEME (Société d'Etudes Mécaniques et Energetiques) and SIGMA (Société Industrielle Général de Mécaniques Appliquées), though both acquired a number of licensees. Between 1948 and 1960 the free piston engine enjoyed increasing custom, evinced by the fact that more than 407 000 hp was in service or on order at the end of the latter year for driving electric generating sets, compressors, oil pipeline and water pumping sets, locomotives and ships. The largest power plant in operation was 35 000 hp and the largest marine installation 16 000 hp in a French naval vessel. Of the 173 units reported in operation at the end of 1960, the principal one was the SIGMA GS 34, a machine with cylinder bore of 34 cm diameter, operating at a boost pressure ratio of 4:1 and imep of about 200 lbf/in^2.

At the centre of the free piston engine is a two-stroke diesel cylinder with two opposed pistons, each rigidly connected to a compressor piston of substantially larger bore. The diesel cylinder is surrounded by a cylindrical case which forms the frame for the complete unit and is also the receiver for scavenged air. At each end of the case is a compressor cylinder; there are no connecting rods or crankshafts, the only physical link between the two opposed pistons being a swinging mechanism which serves merely to maintain piston synchronisation and operate or control such accessories as the fuel injection pumps and the lubricating oil pump.

What happens in operation is that the pistons alternately bounce from one end of their stroke to the other. As they move inwards the air trapped between them is compressed to higher than typical diesel pressures, and fuel is injected normally to promote combustion, the pressure of which drives the pistons outwards. As they move out, the space between each compressor piston and the end of its chamber is reduced and the air trapped in each chamber is compressed until equilibrium is reached and the pistons stop. The air then expands and drives the pistons inward again.

Each air compressor cylinder is fitted with a large number of air inlet valves in the periphery, and air transfer valves through which the air displaced on the compression stroke may be transferred to the scavenge air receiver and thence passed through the ports in the cylinder walls into the power cylinder. In the GS 34 this scavenge air is pressurised to about 3·2 atmospheres and discharges into the scavenge air receiver at a temperature of about 177°C – but it is there heated to about 218°C because blow-back of

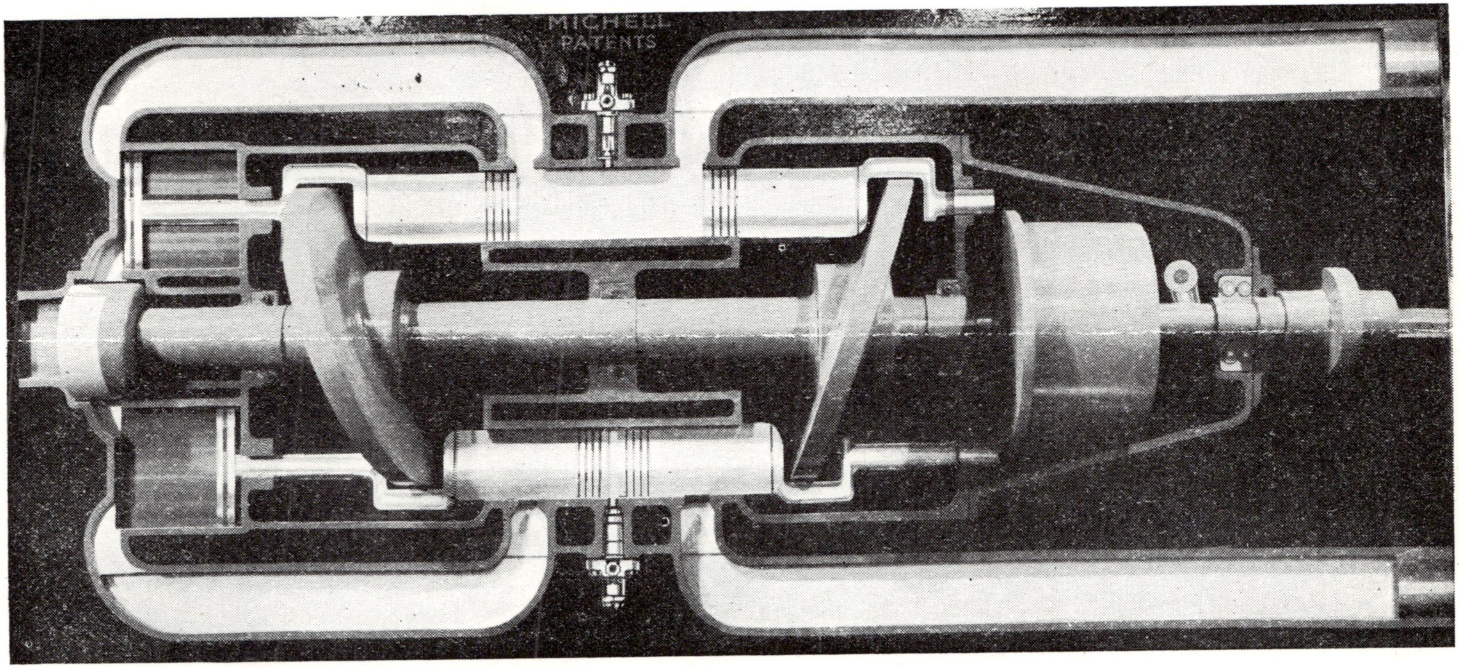

The Sherman-designed Sterling swashplate engine of 1937 was a true 'swashplate' engine, of the opposed-piston two-stroke diesel type, which appears to have given satisfactory service for heavy-duty stationary plant work. Licenced under Michell patents it was also considered, in lighter form, for aero use, but production was confined to the stationary type.

Courtesy: Dr E S Moult, CBE.

Roll out the barrel: The Herrmann experimental aero engine developed at the beginning of World War 2 was virtually two engines in one, consisting of two barrel-formation blocks each containing six parallel open-ended cylinders. Both barrels were joined together by a bolted flange, with their inner ends forming a chamber for a double-sided cam plate which formed two complete sine waves. Two pistons of opposite co-axial cylinders were made as one, with the solid web between them supporting two rollers which made contact with the opposite sides of the cam plate, the rollers being carried on straddle-mounted piston pins employing half journal bearings. The valve gear is also worthy of mention, since it also employed the face-cam principle. At the end of each barrel was mounted a cylinder head carrying inclined inlet and exhaust valves and this inclination brought the tips of the valves near the revolving centre shaft which carried a bevel-faced two-lobe cam plate which acted directly upon them. The inner lobe took care of the exhaust valves, and the outer one coped with the inlet cycle.

Courtesy: Aerosphere *1941.*

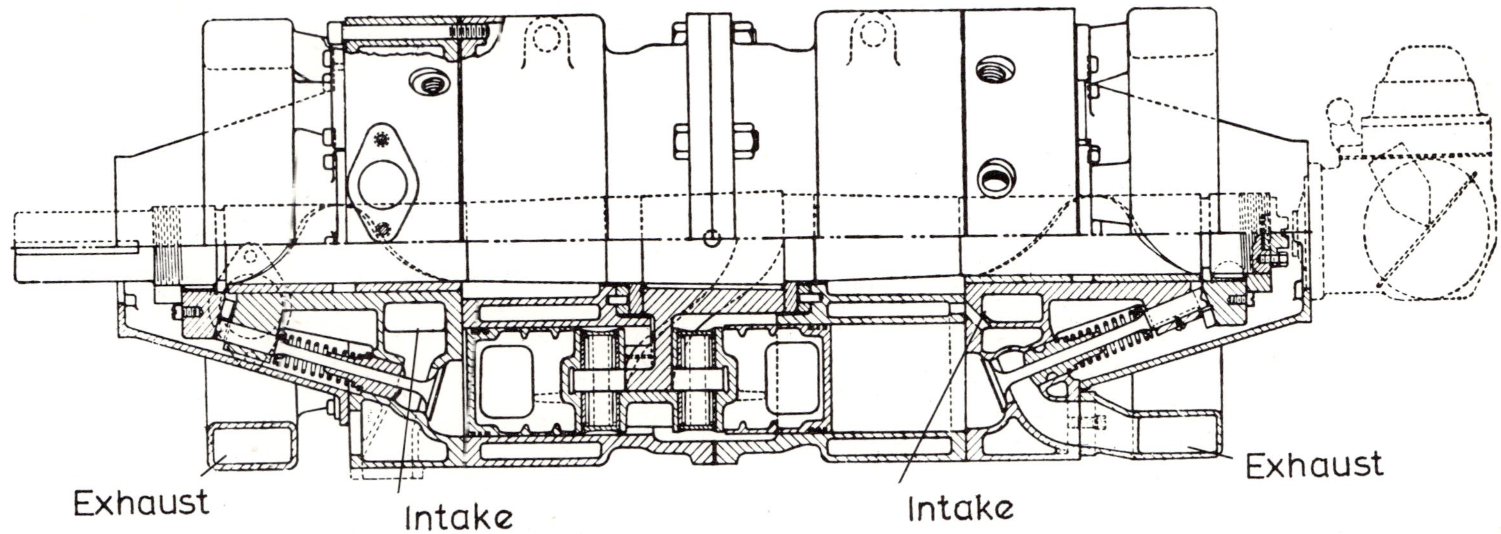
Exhaust
Intake
Intake
Exhaust

combustion gas through the scavenging ports, and heat flow from the power cylinder itself, raise its temperature and pressure further.

At the time of introduction of fuel in the combustion chamber, pressure has risen to about 21 atmospheres; the final firing pressure reached when the pistons come to a stop at the inward limit may be as much as 140 atmospheres or even more. The gas temperature is correspondingly very high.

The free piston engine promises to be difficult to control, but the difficulties are not the most obvious ones. Power output as usual is controlled by varying the amount of fuel injected. The cylic speed or rate of bounce is nearly constant, but the length of stroke for each piston changes somewhat, being shorter at light loads and longer at full load. Fuel injection can be controlled by any one of several conventional arrangements to suit the load characteristics that the turbine may be required to carry.

Real difficulties come in starting and stopping. The engine has to be started with the pistons at the extreme outer limits of their travel, when compressed air at about 21 atmospheres is supplied suddenly to each bounce chamber, driving the pistons inwards at high speed. Because there is no flywheel effect to carry the pistons through several cycles, the engine must fire on the first stroke, so the fuel system including the injection pumps and nozzles has to be completely primed with an oil that will guarantee reliable starting when the engine unit is cold.

An unique feature of the free piston engine is that its mechanical operating parameters (that is to say, its stroke and frequency) are directly linked to its thermodynamic parameters, which are fuel input, delivery pressure, bounce pressure and so on. Furthermore, the combination of a variable inner dead point and very high acceleration rates in that region creates unique problems in fuel injection and combustion. There have also been some quite unusual problems in matching the gasifier to its turbine, especially in installations where more than one gasifier feeds a given turbine. The substantial pressure pulsations in the interconnected exhaust systems create some real headaches. Partial admission turbines were tried in order to maintain gas pressure at sufficiently high levels under part-load operation, and a lot was learned about exhaust phenomena in the process.

The most serious problems, though, were in connection with what Professor Wallace called operating stability. This embraces the effect on the stroke and frequency of sudden changes in load or fuel input, and cyclic irregularities as a result of external pressure pulsations and the like. Starting and stopping the engine could be seen as limiting cases for the former type of instability, but these problems in general were numerous and were probably responsible for the eventual abandonment of most free piston schemes.

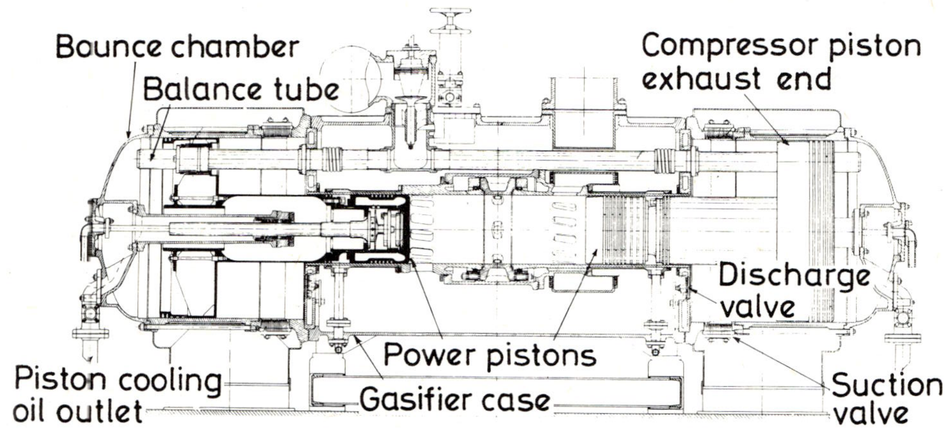

1960 SIGMA free piston GS-34 Gasifier. Commencing with the pioneer work of R P Pescara of France in 1922, development work continued uninterrupted—with the exception of the war years—and by 1960, units of 35 000 hp were in operation, the largest marine installation producing 16 000 hp in a French naval vessel.

Courtesy: Mobil Oil Company Inc.

Some of the same problems may be remarked in other engines relying on the compounded turbine principle. In its simplest form, as exemplified by the dramatically successful Wright Turbo Compound aero engine, there is no problem at all; when the designer gets too clever, or too greedy, and embarks upon a course of further complication, interconnection, and indeed of frustration of the simple piston engine's basic character, it is too easy to end up with a situation in which it is difficult to decide which element is driving what. When they are all working, there you are; but when none of them is budging, where are you? The only safe conclusion to be drawn is the one arrived at independently by Bristol and by Fiat in the late 1930s, when they were pursuing their studies of compounded engines and turbines: once the point has been reached where the exhaust turbine develops so much power that the piston engine merely feeds it and no longer contributes to output shaft horsepower, it is better to go the whole way and invent the modern gas turbine. It may be entirely different, and not what was originally intended; but at least nobody nowadays could criticise it for being unusual.

6. Future Conventional?

The future, according to Thornton Wilder, *is the most expensive luxury in the world.* Certainly no engineer could afford any but the most noncommittal interest in something so deplorably unpredictable. Unfortunately – and perhaps unfairly – the engine designer is stuck with the future, whether he cares for it or not; even now, as environmentalists bellow to conservationists like mastodons across the primeval swamps, as politicians cast about for the bandwaggon offering the fastest and most comfortable ride to fame and fortune, and as some yet mute and inglorious American lawyer is doubtless preparing to inherit the mantle of Nader, the mass of people outside technology promise to make an unholy mess of all the good done for them by the tiny minority who have been schooled in it. For too many centuries it has been thought good sport to see the engineer hoist with his own petard; and if all the reformers had their way it would be a very unusual engine indeed that survived. The hideous inconsistencies spawned by the Clean Air Acts of the USA should constitute sufficient evidence that the only people competent to remedy the errors committed by technology are technologists; but there seems very little hope of the legislators accepting this proposition.

Probably the only hope for the future of the engine is in that very conservatism, that wire-locked and hermetically sealed inertia, which we deplored at the beginning of this book. The public campaigns that wreak their temporal mayhem are essentially manifestations of what is considered fashionable for the time being; and it is of the essence of fashion that it is ephemeral. The popular rallying cry of today may be recollected with embarrassment tomorrow, and all the emotional excursions to various science-confounding extremes may be mutually self-cancelling, having no long-term effect upon the mean course of engine development. It would not be at all surprising to find, twenty years hence, that the engines that we have learned to accept during the hundred or so years since the times of Karl Otto and Sir Dugald Clerk will have continued in normal use, and that all the proposals for unusual engines in the 1970s and 1980s will be granted the same kindly indulgence, as lost causes, as the majority of the earlier unusual engines already reviewed in this book.

If Dean Inge of St Paul's was right in saying that our chief interest in the past is as a guide to the future, then in remembering the unusual engines that have been we might be well advised to forget about the unusual engines that may yet be. Indeed, if we review the century since Clerk and Otto, we may conclude that the only really successful unusual engine to emerge in that time was the compression ignition engine; and after Dr Diesel's original compressed-air injection of pre-heated fuel had been vandalised by the proponents of Ackroyd-Stuart's airless injection, the widespread adoption of the compression ignition engine may be thought to have done more harm than good. That fictional 'average man' of the law courts, the man on the Clapham omnibus, is doubtless unaware that an unusual engine is making his ride slower, rougher, noisier, and dearer than it should be; but his is the ignorance of which bliss is born.

Nevertheless it is as difficult as ever to resist the temptation to folly. Shall we seek to be wise, then, and predict a future in which the patterns of usage of the internal combustion engine will grow significantly different from those to which we are accustomed? Shall we take our heads out of the sand and recognise that short-term and long-term energy crises may materially shorten the engine's expectation of life, significantly reduce its power, strictly confine its opportunities, and hopefully increase its efficiency? If we do, we must examine in baffling detail a forbidding number of factors that are too little known to permit any conclusions to be drawn. If we do not, we might as well forget about it all and take up horse breeding or nuclear physics, in either of which studies may repose the hopes of tomorrow's civilisation – if there is to be any.

If we suppose that our engine designers and manufacturers will henceforth be subjected to disciplines reflecting a sensible compromise between the conflicting demands of fuel economy and environmental probity, we can easily go on to predict the usages whereby any newcomer to the ranks of unusual engines will be inhibited. Where the piston engine survives, it will pursue two alternative courses, either to raise the performance of heavy goods vehicles or depress that of private cars. Locomotives should tend towards electrification, ships to nuclear power or gas turbines according to size and duty. The motorcycle will be proscribed because it is a source of pleasure and an expression of free individuality, neither feature being likely to appeal to the totalitarian governments that we are destined to suffer. The *status quo* will remain virtually undisturbed in one class of engines, however: since we and even totalitarian governors must eat, the engines of agricultural machinery are likely to be ignored.

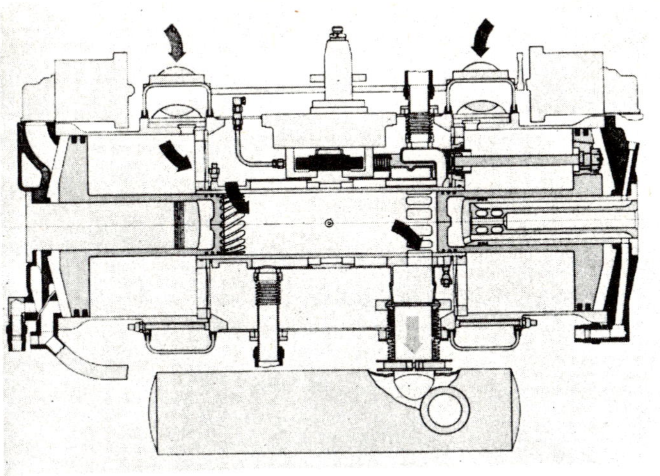

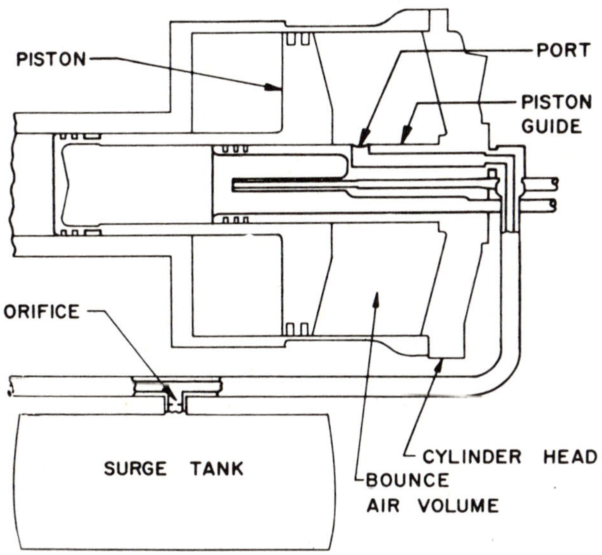

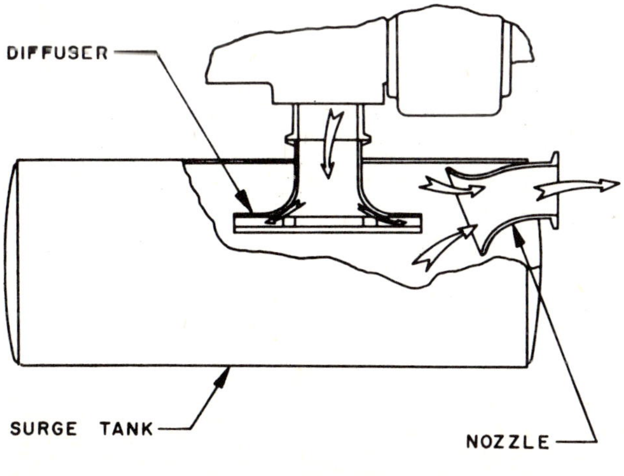

The Ford Motor Company commenced a programme of free piston engine development in February 1954 with a view to the designing of an industrial or automobile/commercial vehicle power plant. The main purpose of the preliminary research was to secure thermodynamic and design information In this the project—designated 519—was successful, but no volume production was undertaken. Initial difficulties encountered and overcome included short piston ring life and high flow losses from the gasifier to the turbine. These were respectively remedied by fitting chrome-plated steel rings in the power cylinder and by the employment of a flat-plate diffuser in the surge tank inlet and a rounded edge nozzle in the outlet.

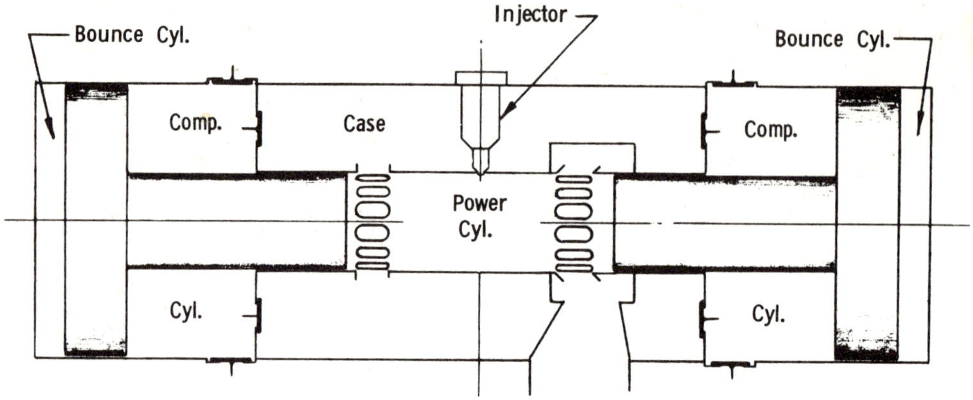

The GM 'Hyprex' engine. Idealised section showing pistons in starting position. Intake and exhaust ports open. All valves closed.

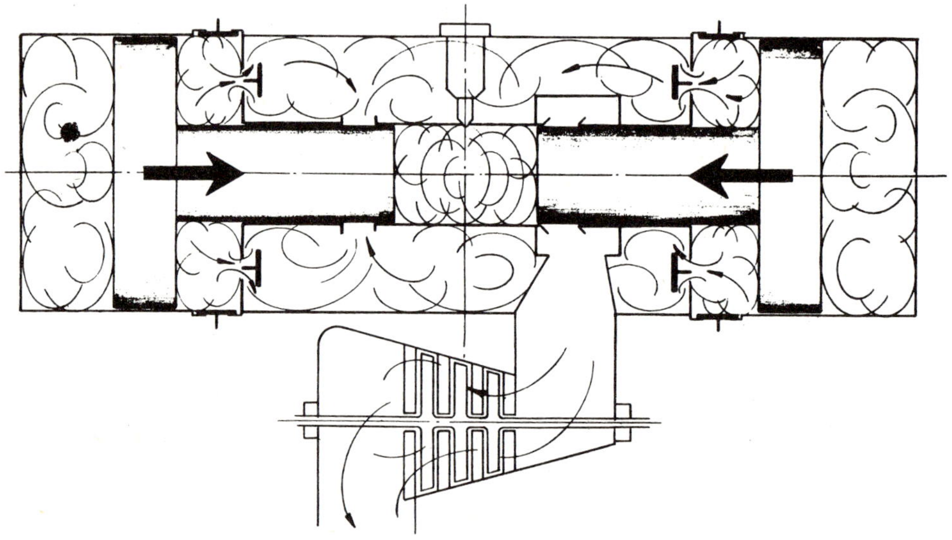

Starting air pressure admitted to bounce cylinders. Pistons move inward, closing ports, compressing air in power cylinder, and forcing scavenging air from compressor cylinder into case.

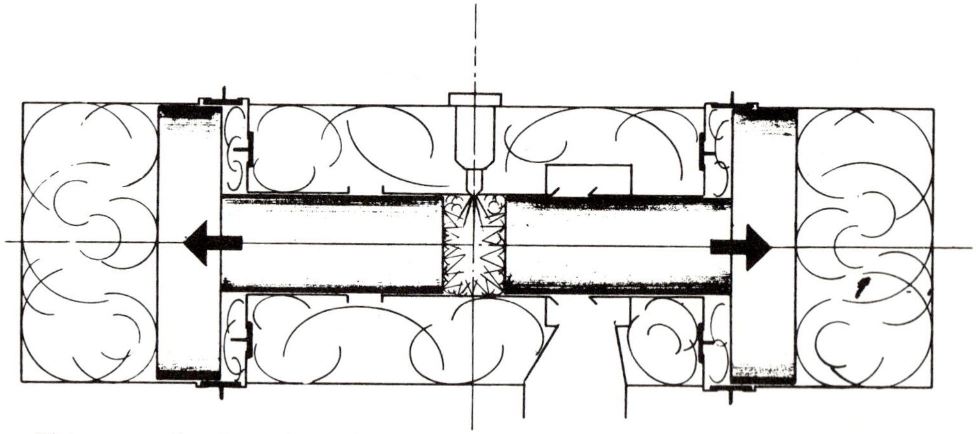

Pistons complete inward travel. Fuel is injected into power cylinder. Combustion starts and power stroke begins.

Courtesy: General Motors Corporation

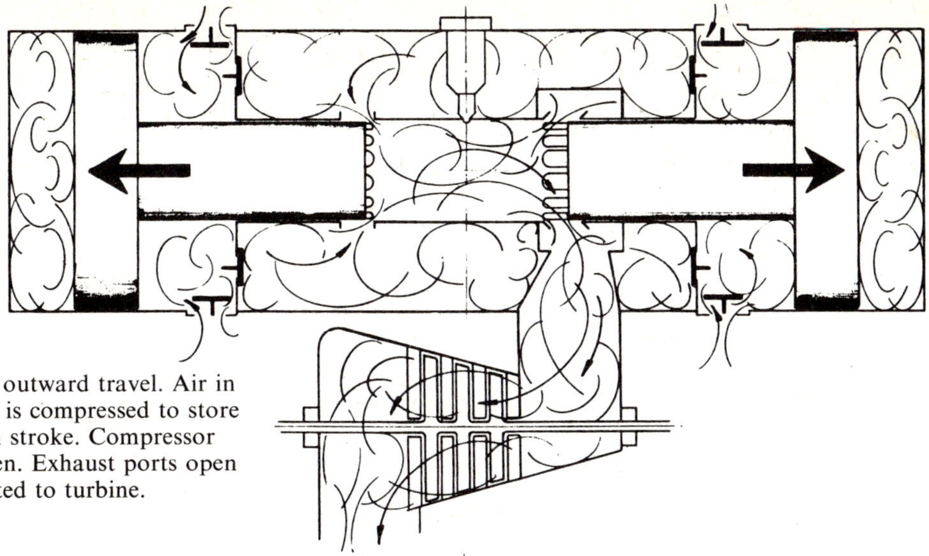

Pistons continue outward travel. Air in bounce cylinders is compressed to store energy for return stroke. Compressor intake valves open. Exhaust ports open and gas is admitted to turbine.

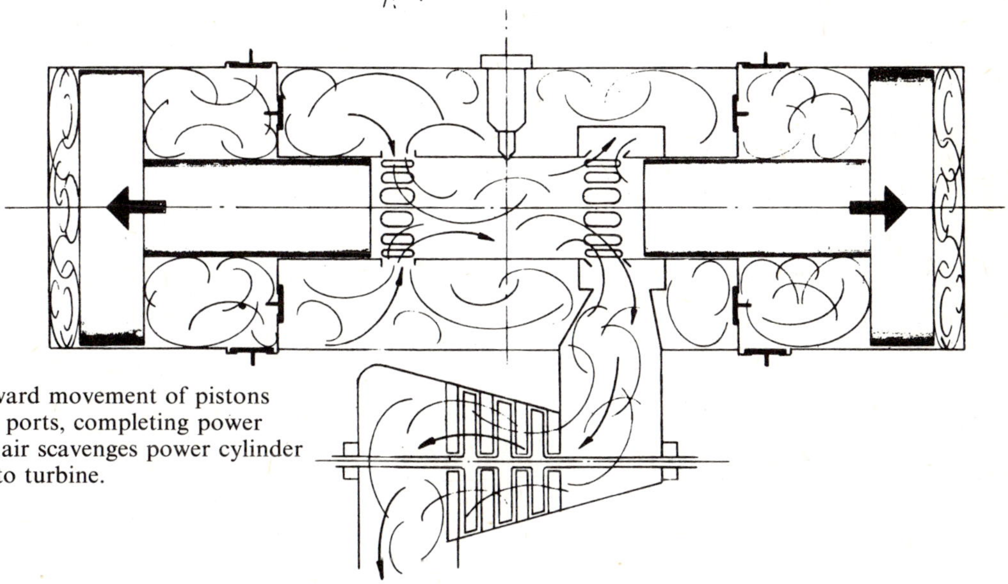

Further outward movement of pistons opens intake ports, completing power stroke. Case air scavenges power cylinder and escapes to turbine.

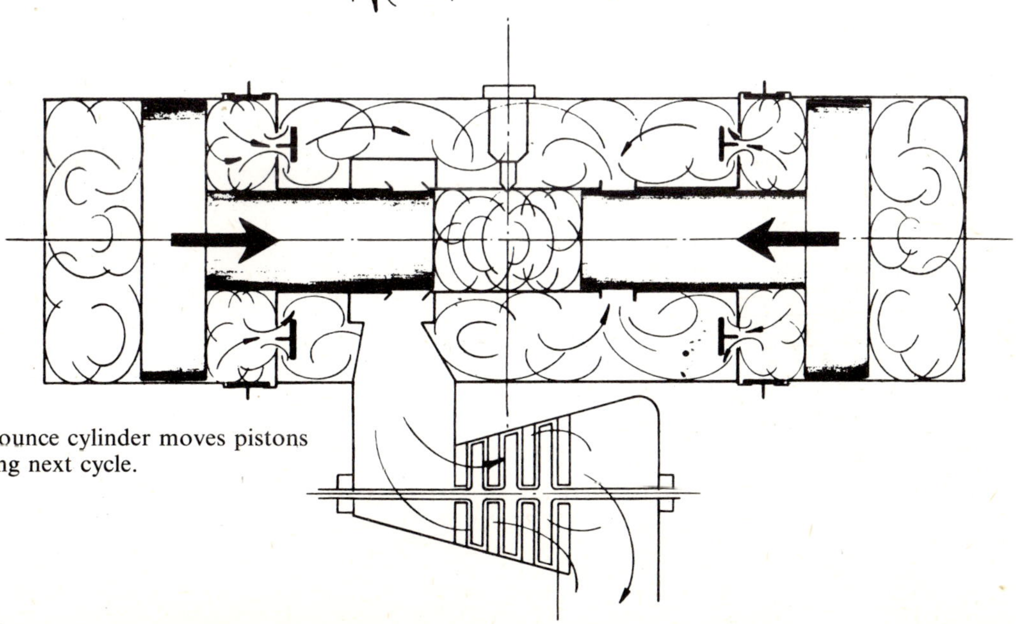

Pressure in bounce cylinder moves pistons inward starting next cycle.

While Ford successfully mounted their free-piston engine in an agricultural tractor, General Motors went one better by installing their GMR 4-4 Hyprex twin cylinder unit in their experimental XP-500 car. This was the first 'Siamesed' unit ever to be built—the configuration being chosen to secure compactness allied to light weight, improved thermal efficiency and a reduction in general noise levels. Like other larger units already in use for stationary applications, the GM engine was basically a uniflow opposed-piston two-stroke diesel with a built-in air piston compressor for scavenging purposes. A simple linkage—usually a parallelogram—kept the two power pistons in each cylinder in phase.

The speed of the engine was controlled by adjusting the amount of air in the 'bounce' chambers by means of a two-way air valve or stabiliser which permitted air to be transferred from the chambers to the air box or vice versa. By removing air from the bounce chamber, the engine was slowed down. In a specially designed chassis, the left hand side frame member housed the insulated pipe connecting the gasifier to the turbine located on the back axle. This permitted a low floor in the vehicle.

Courtesy: General Motors Corporation.

As the panic over noxious emissions gradually gives way to the panic over dwindling supplies of fossil fuel, the aspect of engine design least likely to be ignored in the future is thermal efficiency. Already nearly a third of the oil the world has burned has been wasted in heating the atmosphere. This wastage is likely to be proved to have been unnecessary, although in the early years of unsatisfactory engine materials it was at least excusable. This passion for dissipation of heat went hand in hand with a conviction that not only should the engine's components (especially the cylinders, pistons, and head) be run as cool as possible, even by exploiting inherently unsuitable metals for the sake of their superior thermal conductivity, but also the utmost pains should be taken to introduce the charge of fuel and air at the lowest feasible temperatures for the sake of high volumetric efficiency.

Now it seems beyond contradiction that if you introduce cold air and cold fuel into a cold engine you will experience some difficulty in getting and keeping them hot; and the internal combustion device is supposed to be a heat engine. Practically all the problems of polluted exhaust, notably from unburned hydrocarbons and from aldehydes frozen in the middle of the chain reaction that goes on in the flame front during combustion, may be attributed to the excessive quenching of a charge burning at too low temperature and pressure to survive being brought into proximity with over-cooled cylinder walls, piston crowns and cylinder heads.

This at least is the conviction of that unusual engineer P. C. Vincent, best known for his eponymous and admirable motorcycles of a bygone and fondly remembered age. He has some cause to think along these lines, for in the 1960s he embarked upon the design and construction of a modern high-speed rotary engine, in which the rotating annulus on the outside was a sort of continuous-process cylinder head carrying all the necessary porting, provisions for ignition, and so forth, while the crankshaft at the centre of the drum-shaped block containing diametral cylinders would either remain stationary for two-stroke engines or rotate in the opposite sense to the head for four-strokes. The original object of this layout had been to provide a system of induction and exhaust timing that would automatically attune itself to the speed of the engine, unlike the conventional poppets or ports which favour volumetric efficiency at a pre-determined (with any luck!) engine speed to the detriment of breathing at other speeds.

Since the engine was expected to combine high volumetric efficiency with high rates of rotation, from which a considerable heat throughput might be predicted, Vincent took the trouble to design it with exceptionally generous cooling provisions – and was rewarded for his trouble with an engine so resolutely cool that it was unable to develop enough heat to warm up the piston crown, the head or the cylinder liners, enough to avoid serious and permanent quenching as the pistons rose to top dead centre. In practice the engine stopped as soon as the electric starting motor was switched off, even with no load applied to the brake; the only measurable output was a black and viscid stream of mixed oil and petrol trickling from the exhaust pipes.

The orderly substitution of iron for aluminium alloys in one component after another, together with other steps to constrict the paths whereby heat might escape from the engine, successively made it able to sustain itself and eventually to develop a modest but satisfactory power output.

This corresponded to 22·4 hp/l; and whilst it is undeniably a great deal better than zero net output, it hardly commends itself as an appropriate rating for an engine boasting volumetric efficiency of 120 per cent. What checked Vincent in his extrapolation of the improvements he wrought was the predictable onset of pre-ignition caused by the high surface temperatures now being realised within the combustion space.

We have seen that there are two ways of preventing pre-ignition from being a problem. One is to make a stratified charge engine, the other is to make a Diesel. Vincent was attracted to the latter; but his engine was to be a petrol-burner, not an oiler, and in fact the solution he hit upon had something of the characteristics both of the Diesel and of the Honda stratified-charge CVCC engine described in Chapter 1. The outstanding feature of his new design is the pressurised introduction of fuel into a small ante-chamber, in which it is heated to the temperature at which it will ignite spontaneously, before it can come into contact with any air. This presupposes the manufacture of this swirl chamber from one of those strong new synthetic ceramics, which ally good strength with very low heat conductivity and remarkable ability to resist extremely high temperatures without oxidising or similar damage being suffered. Several of these synthetic ceramics act as good electrical insulators, so the swirl chamber made from one of them could simply be warmed by electric heating elements wound around it to pre-heat it before starting from cold. There is some evidence to indicate that the inflammability of hydrocarbon fuels increases so much with increase of temperature that no more than 300°C need be reached to ensure combustion for starting.

With this pre-heating of the fuel, the very high compression ratios characteristic of the compression ignition engine can safely be avoided, to the benefit of stress levels and pollutant levels alike. As the piston rises to top dead centre, it forces air into the ante-chamber through small communicating orifices. Even if the compression ratio were only 5:1, this would promote a velocity into the chamber about 9/10ths of that given by a 200:1 compression,

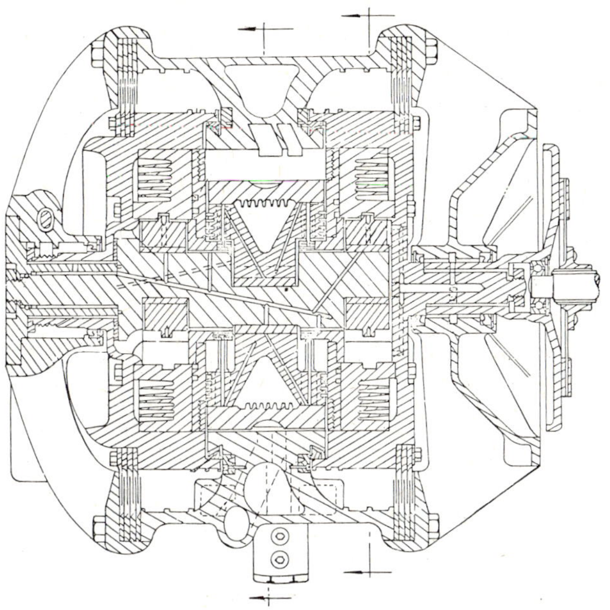

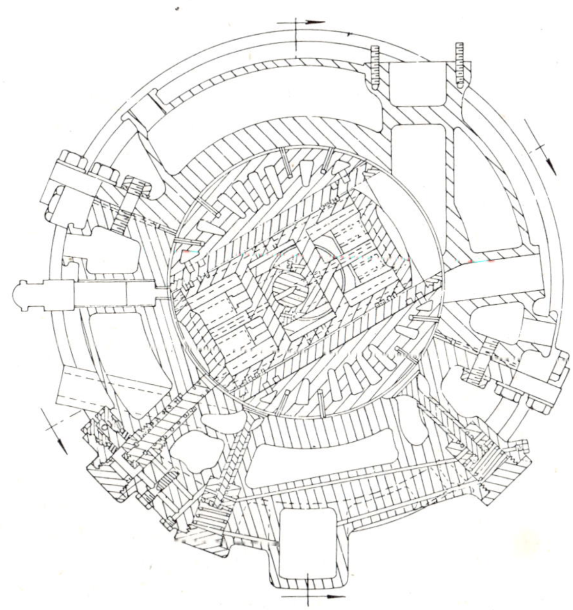

One might expect a design by Philip Vincent to be as iconoclastic as his writings, and in this experimental engine he packs (within the compact confines of a continuous annular cylinder head) a great deal that is both literally and metaphorically revolutionary. Features to note include a Scotch yoke in the short-stroke piston trunk, a profusion of seals and oil galleries, and provision for Vincent's hot fuel injection/ignition chamber and nozzle.

and it is heated on its way through the passage even when the engine is cool. The introduction of air to the heated fuel will produce ignition and the resultant increase in pressure will force a mixture of burning fuel and air under pressure through the orifices and back into the cylinder full of air, thus restoring to the engine the original hot compressed-air injection invented by Dr Diesel.

The throttling of the passage between the ante-chamber and the cylinder is done by a valve made again from ceramics capable of resisting without damage high temperatures that could burn even the latest and best Nimonic alloys. This valve is not subject to any violent accelerations, no matter how high may be the rate of crankshaft rotation, its only motion being a slow and gentle linear one, which varies the degree of throttling of the orifices and thus effects control of advance and retard of ignition timing.

In Vincent's design there is no provision for any cooling of the ignition chamber, only the inlet charge of air serving to prevent it from becoming excessively hot; this air will be relatively very cold as compared to the hot ceramic walls, and so it will take up heat from them before ignition, in turn aiding rapid combustion. The combustion itself is expected to be relatively cool, in part due to the modest compression ratio and in part to the very rich mixture present. Only a small part of the fuel charge can be burned in the ante-chamber because of the limited quantity of air, the remainder of the fuel present simply being heated there, and not burning until combustion pressure in the ante-chamber forces it through the cylinder.

In principle, therefore, the temperatures and pressures reached during the combustion stage should not reach the levels at which oxides of nitrogen are likely to be formed. Moreover, since the impossibility of pre-ignition allows the engine to operate on low-octane petrol, lead and all other pre-ignition inhibitors should be unnecessary and need therefore not sully the exhaust.

All this will only be true if the combustion proves to be truly resistant to the slightest sign of pre-ignition – if, in other words, the throttling valve can effectively control the speed of injection of the fuel to the cylinder after ignition, thus preventing too rapid a pressure rise in the cylinder. Doubtless there will be enthusiasts for the constant-pressure cycle who will welcome the late introduction of fuel into the cylinder so as to keep pressure down, but equally doubtless those who accept the advantages of the constant-volume cycle will lament that the full potential for the expansion phase to take place through as much of the piston stroke as possible cannot be enjoyed in the Vincent engine. Nevertheless its designer foresses an overall thermal efficiency in the region of 70 per cent since the free use of ceramics (most probably silicon nitride) will allow very high temperatures to be reached and encourage

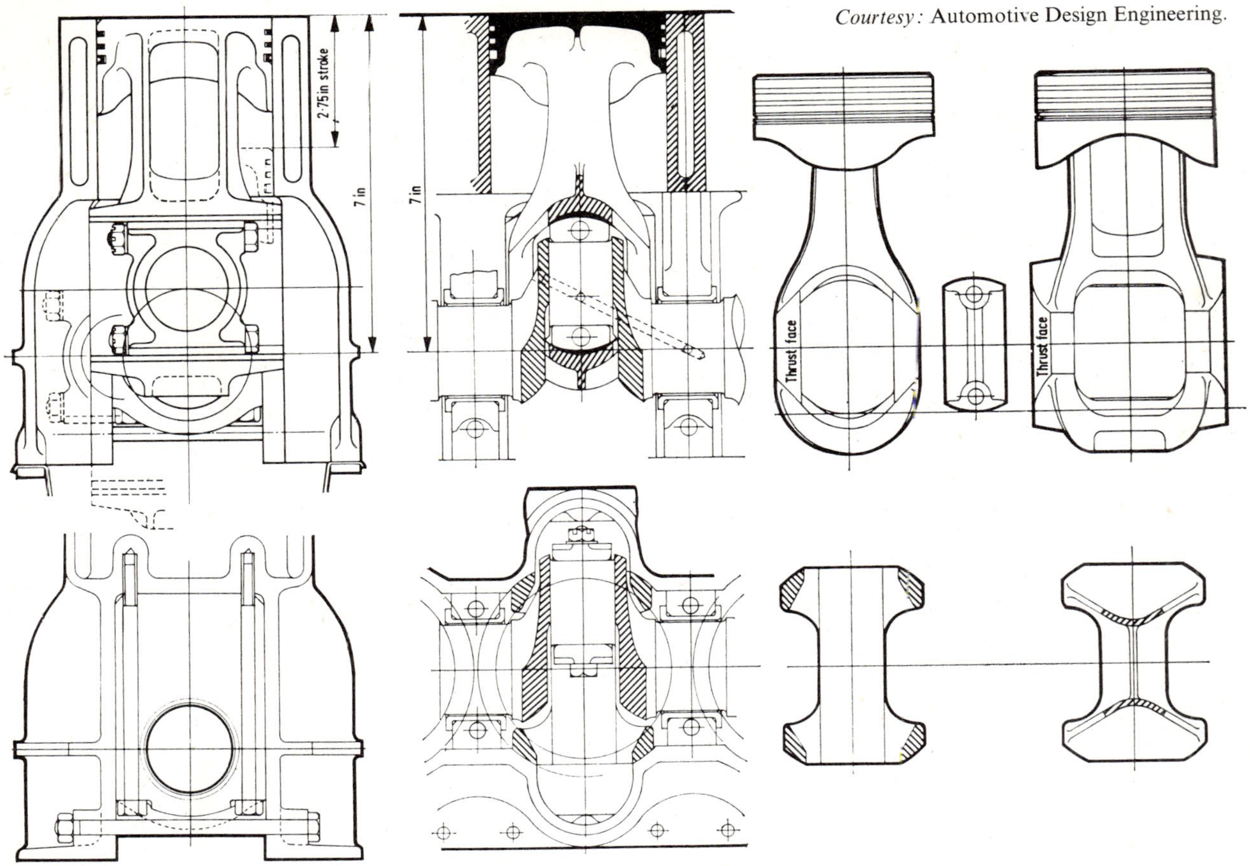

Courtesy: Automotive Design Engineering.

Set out in a theoretical paper on the piston engine by the late S S Tresilian and published posthumously, the Tresilian 'Scotch Yoke' principle illustrated here employed a system which has been tried with varying degrees of success in the past. The Knight split sleeve-valve engine designed for the K-D car (Knight-Davidson) in 1913 (no connection with Charles Yale Knight's double sleeve valve designs) by M E Knight, employed 'scotch yokes' but these were later replaced with conventional connecting rods and big ends.

the most parsimonious cooling system imaginable. In fact he is talking of raising the fuel and air charge ultimately to over 1000°C, ensuring instant reliable ignition spontaneously without high compression ratio.

Much will depend on whether his ceramics can stand it. There is no doubt but that they have the thermal characteristics he claims, but it is less certain that they can be preserved in his design from the tensile stresses that will shatter them, nor that they will survive without lubrication in those places (such as the piston) where they are subjected to friction. It has been reported by powder metallurgists in the BSA group that unlubricated surfaces of polished reaction-sintered silicon nitride have a mutual coefficient of friction of 0·1 when sliding under normal loads up to 75 lbf/in^2 – but even the Scotch yoke mechanism, as adopted by Vincent for the linking of the piston to the crankshaft, seems unlikely to achieve the reduction in bearing pressures necessary to keep within these bounds.

Perhaps in time we really shall have an engine made entirely of ceramics, whether it be Vincent's or some other. Perhaps in time an engine embodying the Scotch yoke will go into production. Certainly there are plenty of engineers who would like to see the latter, for the Scotch yoke has attracted several of the most outstanding brains in the business, not least of them being the late Stuart Tresilian.

The beauty of this mechanism, which is already used in reciprocating oil pumps, is that it renders otiose the articulated connecting rod; in turn this produces true harmonic motion for the piston, with reduced accelerations, due to the elimination of the troublesome angular swing of the connecting rod. What is more, it allows an engine of any configuration to be balanced without regard for secondary forces. Best of all, it liberates the engine designer from all sorts of unwelcome constraints in the determination of stroke: bore ratio, connecting rod length, piston side thrust, gudgeon pin location, and the height of the cylinder block above the crankshaft centre line. It is a particularly attractive idea when embodied in an opposed-cylinder layout and in engines of ultra-short stroke. Of course, all these claims could be made of the axial engine with cam plate, such as the Herrmann described in Chapter 4 – and, looking back, the dissemination of that particular gospel could hardly be likened to wildfire. Whether the theoretical advantages of the Scotch yoke suffice to outweigh the probable penalties in mechanical friction from the block that slides in the piston-skirt crossways (a location that is hardly conducive to efficient lubrication) had better be left for empirical determination; but it was made clear to the Institution by Hassan and Mundy (see chapter 2) that one of the best remaining ways of gaining power is to lose friction.

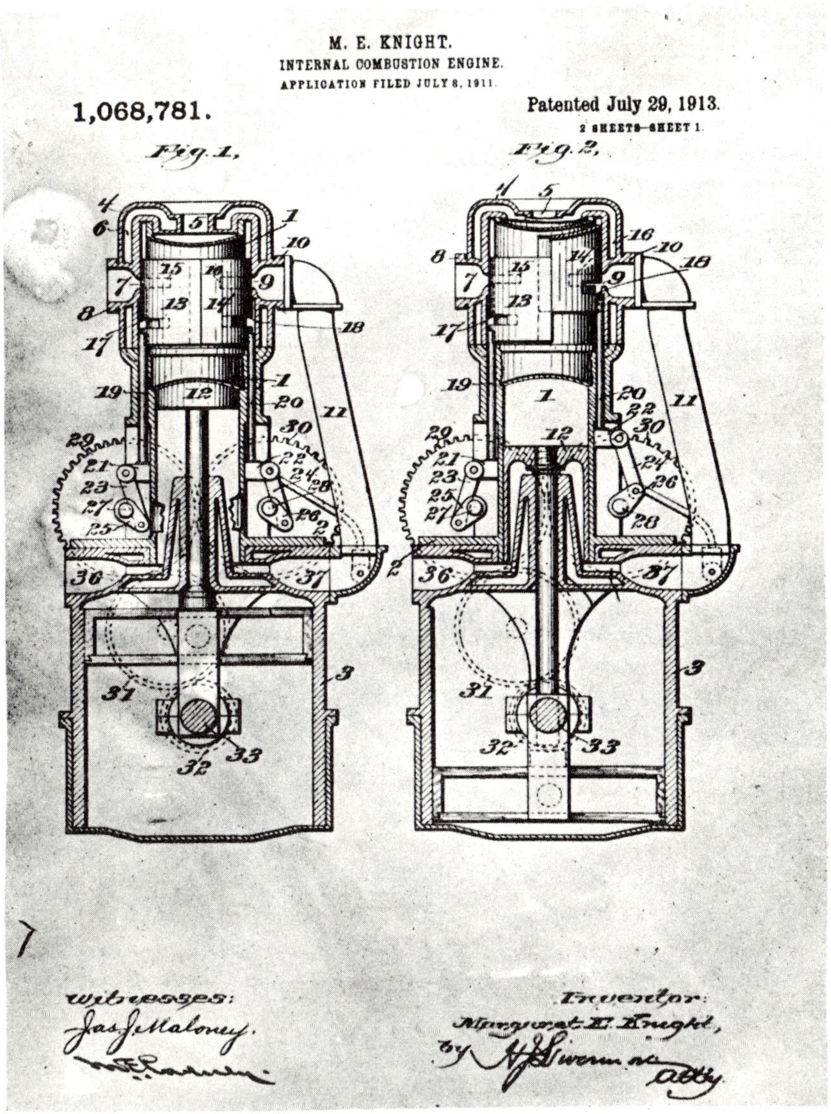

Drawing from patent for sleeve-valve engine issued to Margaret E Knight, assigned to the K-D Motor Company.

Courtesy: G Marshall Naul and Automobilists of the Upper Hudson Valley

One of the traditional ways was to run the engine at higher speeds than might have been thought customary in the past. Those accomplished miniaturists, the Japanese, have certainly shown us the way: the exquisite little 5 cylinder 125 cm^3 racing Honda, that sang its shrill way up into the regions beyond 20 000 rev/min, is probably the best example yet released. Nevertheless, there are higher-revving engines in Japan than ever came ever out of it: in the late 1960s they were conducting research into engines running at rates in excess of 50 000 rev/min. The idea is not without its attractions: when cyclic frequencies are so extraordinarily high, it may be possible to achieve by means of shock waves generated within the cylinder not only means of sealing the bore without friction-inducing piston rings, but also perhaps even of throttling intake and/or exhaust ports without having to rely on mechanically operated valves. There is perhaps a danger that starting such an engine would require a motor and gearbox bigger than the engine itself . . .

As an alternative to the very-high-revving four-stroke of which the racing Honda motorcycle is such an exemplar, the highly tuned two-stroke has enjoyed the most amazing development during the past decade. To achieve a specific output in excess of 300 bhp per litre from a little 2-cylinder two-stroke of absurd simplicity is a feat guaranteed to make any power-hungry racing man fall in love with the two-stroke – so long as he can afford its fuel consumption. But if 'when poverty comes in at the door, love flies out through the window', then it is clear that when volumetric efficiency comes into two-stroke design, thermal efficiency dives out through the exhaust port. The rabid thirst of the currently fashionable high-output two-strokes employed in motorcycles, snowmobiles, and target drone aircraft, is hardly likely to endear the type to the rising generation of sophisters, economists, and calculators. Yet the glory of the two-stroke may not necessarily be extinguished; it may have been a wise and salutary neglect that suffered it to find its own way to perfection, for only now has the technology developed whereby the once humble two-stroke engine may be transformed by the processes of mechanical sophistication to become a power unit of considerable promise for the future. The engine has a lot more in its favour than many may suppose; and any manufacturer who is not a fully paid-up member of the Wankel Club might do well to break down the barriers of superstitious nonsense that smother the brilliance of the unnecessarily humble two-stroke cycle.

This may sound like the preaching of heresy; but, to the true believer, the greatest heresy of all is that the two-stroke is bereft of all virtues except mechanical simplicity and low weight, and fit for nothing but such applications as motorcycles and lawnmowers where crudity and thirst are not automatic disqualifications to service. Allow the two-stroke as much mechanical complexity as a four-stroke, give it uniflow scavenging and a mechanically operated exhaust valve, and direct fuel injection into the cylinder, and positively pressurised induction – ideally with compound turbo charging – and it can give a very good account of itself in terms of power to weight, power to displacement, power to cost, or specific fuel consumption ratios. What with its low internal pressures and the economies of direct injection (not least of which is the benefit which comes from scavenging with pure air) it offers a tempting means of abating the nuisances of exhaust pollution and fuel depletion.

Many are the engineers who have dreamed such dreams in the past, no doubt. They were but pipe dreams because in the past we were not particularly concerned by such problems as environmental sanitation or fossil fuel consumption. At long and perhaps overdue last we are forced to take such matters into account; the times and the two-stroke are ripe for exploitation. In the late 1970s we may expect to see the dream debouch from its pipe in the form of a very efficient and practical two-stroke conceived by Audi and nurtured by Volkswagen. It is already giving good results as a prototype. Its specific fuel consumption is down to 0·4 lb/hp h, which is already better than some compression-ignition engines can manage. Its injection apparatus is no more troublesome than is proper to the developed version of something that worked well in the M196 Grand Prix Mercedes-Benz engine, a four-stroke which turned at up to 10 000 rev/min when technology was nineteen years younger. It is quite happy with the latest modes of ignition, the latest kinds of petrol, the latest measures of emission control, the latest mores of economy. It would doubtless be flattered by the emulation likely to follow it in Britain, where another motorcar manufacturer is understood to be working along similar lines.

Many others are turning in their despair to the turbocharger. It may prove their salvation, as it has in other applications; but if it has any remarkable facet in its character, it is that it can do as much for a good engine as for a bad one. From this it follows that those who take it up when their backs are to the wall are likely to find it no more than a palliative, whereas those who adopt it with their minds to the future may exploit it as a means of going on to better things.

The interesting thing about all these various measures now suing for more general adoption – stratification, turbocharging, the two stroke cycle, ultra-high revs, high thermal efficiency, high volumetric efficiency, and the deployment of new engineering materials – are all capable of being embodied in one of the unusual engines discussed earlier in this book. That one – and on the available evidence it seems to be the only one – is the Aspin.

It may take an unjustifiable amount of optimism to expect this unique potential to be realised, but the possibility is there because the engine has always been subjected to continuous review as well as almost continuous development. It is not, in other words, something that would have to be revived, as would any of the other lost causes we have mentioned. *Revivals are shallow things, since they aim at reproducing what never existed or what has perished with the age that gave it birth* (to quote Dean Inge again); but we have seen enough of history to know that commercial success does not depend primarily upon technical suitability for the market.

That is why so many unusual engines must be classed among the engines that might have been, and why so few may find a place among those that might well be. Study the following list of casualties and ponder how often each loss is balanced by a lost cause that perished with it. Yet there may be a further message in the list for the aspiring maker of some unusual engine perhaps yet to be revealed: *your cause is lost, but in the heart of all right causes is a cause that cannot lose*. The usual engine is very far from perfection and a good way off being even satisfactory. The custom that makes it usual makes it corrupt; and if this fate could be averted by a fresh starter, that would be unusual indeed.

A List of Unusual Engines

A list of unusual engines can never be exhaustive, but it can be more wide-ranging than a book text. If the reader hoped to find some particular unusual engine in this volume, and has so far failed, he may succeed here.

ADAMS-FARWELL	rotary 1903
ALBAN	high-pressure steam, 2 single-acting opposed plungers
ALFARO	axial
ANDREAU	differential-stroke, made by Citroën
ASPIN	rotary valve
AZURE BLUE	stratified charge
BISSCHOP	half-stroke gas engine
BOFORS	gas turbine 1947
BRISTOL	wobbleplate
BRISTOL HYDRA	10-cylinder aero radial
BRISTOL PHOENIX	diesel air-cooled aero radial
BRADSHAW OMEGA	toroidal cylinder
BERTIN	hot-air turbine
BRM	H16 racing car, 1966
BREGUET-BUGATTI	double-double straight-8 aero
BROWN BOVERI	industrial gas turbine 1939
BURLAT	contra-rotary
BURT-MCCULLUM	sleeve valve
BUTTERWORTH	swinging valve
CAMPINI	aero gas turbine 1932
CANTON-UNNE	radial with no master rod
CAPITAINE	forerunner of Diesel 1889
CAYLEY	open-cycle
CHATTERTON SIMPLIC	internally-air-cooled hyper-expansion 2-stroke self-charge
CLARKE	precessive rotor
CLERK	two-stroke gas engine
CROSS	rotary valve

CHRYSLER	multi-bank tank engine
CUDDON-FLETCHER	2-stroke
DAY	crankchamber-scavenge 2-stroke 1891
DE LAVAL	impulse turbine with de Laval nozzle 1889
DEMONT	double-acting rotary
DEVILLE-NEGRE	vane-type rotary 1964
EJC	contra-rotary
ERICSSON	open-cycle air engine 1840
FACCIOLI	opposed-V aero
FIAT AS6	aero double V12 contraprop
FIAT TORAZZA	variable valve timing and lift, intake-throttling valve
FRITZ	rotary 1961
FORD FSC	stratified charge
GILBERT	rotary
GOBRON-BRILLIÉ	opposed-piston single-crank
GRAF UND STIFT	2-stroke oil engine for automobiles
GNOME MONOSOUPAPE	aero rotary, single valve per cylinder
GRAY-DRUMMOND	rotary akin to Wankel 1909
GRIFFIN	6-stroke gas engine 1883
GULDNER	early German diesel 1903
HASELWANDER	fuel injection oil engine 1899
D'HENAIN	rotary aero, cylinders and crankcase in monobloc
HESSELMAN	heavy oil, 1925
HERRMANN	axial, cam-plate
HOLZWARTH	explosion gas turbine, industrial, 1925
HORNSBY ACKROYD	oil engine of Ackroyd Stuart type
HUMPHREYS	constant compression 1952
HVID	Diesel type oil, 1913
ITALA V12	racing 1·1 litre with supplementary scavenge air
ISUZU	rotary combustion in 3-lobe chamber
JAMES	'clap-hands' opposed-pistons
JENBACH	two-stroke oil engine for cars 1952
JENDRASSIK	constant pressure gas turbine 1939
KAL-PAC	rotary-block cruciform cylinders
KARAVODINE	pulsating or explosion gas turbine 1908

KAUERTZ	'clap-hands' precessive pistons
KETTERING	high-compression car engine 1947
KNOTT	eccentric type rotary 1863
KÖRTING	double-acting two-stroke gas engine 1898
LAMB	eccentric type rotary 1842
LORENZEN	hollow-bladed gas turbine 1931
LORIN	piston engine for producing reaction jet 1908
LUCAS	valveless petrol engine for cars 1907
LYCOMING	outsize multibank aero 1945
LYSHOLM	multistage centrifugal gas turbine 1930
MALLORY	rotary vanes
MECH-TECH	Brayton cycle V engine
MERCEDES BENZ M196	desmodromic valves, direct injection, racing cars 1954
MERCEDES BENZ M196/110	similar, with variable-length inlet tracts
MERCER	rotating block
MACOMBER	rotary axial engine for cars
MITSUBISHI 10 ZF	air-cooled compound turbocharged diesel V10 for tanks
MICHELL	axial
NEC	supercharged 2-stroke aero 1910
NAPIER NOMAD	compound turbocharged aero diesel
NAPIER SABRE	sleeve-valve H24 aero
ORION	compact tank engine with turbine output
OTTO-LANGEN	free piston 1875
PERKINS	differential diesel
PESCARA	free piston 1925
PORSCHE	cruciform rotary block
PHILIPS	Stirling cycle
REP	fan-form aero
RIDER	Stirling cycle 1876
RENAULT-RAMBLER	rotary-combustion in 5-lobe housing
RIGEL	opposed-piston, same as Orion (q v)
RAJAKARUNA	rotary 1964
ROLLS-ROYCE CRECY	compounded direct-injection 2-stroke aero 1941
SARICH	rotary 1973
SALMSON	6-stroke aero radial

SALMSON	water-cooled aero radial
SALMSON	aero radial with Canton-Unné (q v) rod gear
SCHLAMANN	stratified charge 1953
SELWOOD	orbital 1964
SIEMENS	contra-rotary
SIGMA	free piston
SKLENAR	rotary aero 1939
STENBERG	Stirling cycle
STILL	combined Diesel and steam cycles
STOLZE	first gas turbine (claimed) 1900
STORK	uniflow-scavenged single-acting 2-stroke 1958
THIOKOL DYNASTAR	cruciform twin-piston 2-stroke with two crankshafts 1963
TSCHUDI	precessing pistons
VARIANT	rotating combustion, 4-lobe housing
VINCENT	rotary
VIRMEL	rotary
WISHART	split-cycle 1970
WOOLER	axial swashplate
WOOLER	rockerbeam motorcycle engine

Index